Studies of Narcosis

Contributors

Leonard L. Firestone
Department of Anesthesiology and Critical Care Medicine,
University of Pittsburgh School of Medicine,
Pittsburgh, Pennsylvania

Robert L. Lipnick
Office of Pesticides and Toxic Substances,
United States Environmental Protection Agency,
Washington, DC

Keith W. Miller
Department of Anesthesia, Massachusetts General Hospital and
Harvard Medical School, Boston, Massachusetts

Peter M. Winter
Department of Anesthesiology and Critical Care Medicine,
University of Pittsburgh School of Medicine,
Pittsburgh, Pennsylvania

Studies of Narcosis

Charles Ernest Overton

Edited by
Robert L. Lipnick

United States Environmental Protection Agency,
Washington, DC

SPRINGER-SCIENCE+BUSINESS
MEDIA, B.V.

WOOD LIBRARY-
MUSEUM OF
ANESTHESIOLOGY

First edition 1991 published under the joint auspices of

Originally published by Chapman and Hall Ltd and the Wood Library-Museum of Anesthesiology in 1991

Typeset in 10/12pt Palatino by EJS Chemical Composition, Bath, Avon

ISBN 978-94-010-5371-6

British Library Cataloguing in Publication Data

Overton, Charles Ernest, *1865–1933*
Studies of narcosis and a contribution to general pharmacology.
1. Narcotics
I. Title II. Lipnick, Robert L. III. Studien uber die Narkose zugleich ein Beitrag zur allgemeiner Pharmakologie. *English*
615.7822

ISBN 978-94-010-5371-6 ISBN 978-94-011-3096-7 (eBook)
DOI 10.1007/978-94-011-3096-7

Library of Congress Cataloguing-in-Publication Data

Available

Contents

Foreword

Charles Ernest Overton's 1901 monograph *Studien über die Narkose* has become a scientific classic in a number of different fields. This book represents the first English translation, and in fact the first translation into any other language, of the original German work. In addition to the edited translation, this volume contains introductory chapters by Keith Miller, Peter Winter and Leonard Firestone and myself.

As editor, I have attempted above all else to ensure that the translation faithfully represents Overton's ideas and data, while making the material readily understandable to the modern scientific reader. This has frequently required that extremely long sentences, common in turn-of-the-century German but considered cumbersome today, be simplified into two or even three sentences. In addition, I have paid particular attention to the correct translation of scientific terms, and I accept complete responsibility for any inaccuracies in this area.

Overton's original contents list included headings and subheadings, but only a fraction of these appear in the original text. For the sake of clarity they have all been included in the body of the translated work. Also included is an index containing all chemicals mentioned in the book, along with their Chemical Abstracts System Registry Numbers for unambiguous identification, a complete list of Overton's publications (Appendix B), and a list of all biographical articles about Overton and articles dealing specifically with analyses of his data (Appendix C).

Although the original book contains no figures, four have been added here. They were either cited in the book by Overton, who referred the reader to the original source for the figures, or obtained from Overton's 1895 publication, and were included for the sake of clarity.

I first became aware of *Studien über die Narkose* in the early 1980s when I saw it cited numerous times in both the recent and older literature.

At the time I was beginning work on the use of quantitative structure–activity relationships (QSAR) in the correlation of toxicity with chemical structure. QSAR is used by the US Environmental Protection Agency's Office of Toxic Substances to estimate the potential hazard posed by the release of industrial organic chemicals into the environment. Since such assessments tend to focus on potential release into rivers and other aquatic systems, Overton's use of aquatic organisms is directly pertinent to this mission. With the support of Dr James H. Gilford, Chief (now retired) of the Office of Toxic Substances' Environmental Effects Branch, I was able to arrange for the English translation of this book through a government contract with SCITRAN Inc., Santa Barbara, California. By the summer of 1985 I had in hand a complete translation of the book, and it was immediately clear to me that it would be of great importance to make an edited version of this translation available to the scientific community since many of Overton's ideas and observations were continually being rediscovered.

In September 1985 I was invited as the result of a recommendation by Professor Roelef Rekker in the Department of Pharmacochemistry at Vrije Universiteit, Amsterdam (now retired) to prepare a leader article for publication in *Trends in Pharmacological Sciences* (*TiPS*) to better inform the scientific community of the importance of this seminal book. This article was published in April 1986, and is reproduced in full in Chapter 3 with the kind permission of the publisher, Elsevier.

Several months after the publication of this article, I was contacted by Dr Leonard Firestone, then with the Department of Anesthesia, Massachusetts General Hospital. Dr Firestone indicated that he had read the *TiPS* article and suggested that the Wood Library Museum (WLM) of Anesthesiology might be interested in supporting the publication of the Overton book translation. He contacted Dr Elliott Miller, also with the Department of Anesthesia, Massachusetts General Hospital, who was then Chairman of the Publication Committee of the WLM; he supported this idea, as did Dr Nicholas Greene, Department of Anesthesiology, Yale University School of Medicine, who is currently Chairman of the WLM Publication Committee, and is the one who eventually brought this book before the WLM Board to request sponsorship and financial support. Dr Greene worked persistently to obtain funds for this project and met on several occasions in London with officials from Chapman and Hall.

During this time I performed the first QSAR study of all of the test data on tadpoles that Overton provided in this book. The analysis clearly demonstrates that although Overton did not have the benefit of modern statistical methods, his reported toxicity values are as reliable as any aquatic toxicity data in the modern literature. This reflects the fact that

Overton was a very careful observer and also that he took great pains to choose highly reproducible end effects.

During the summer of 1989 I was extremely fortunate to receive a grant from the Swedish Crafoord Foundation to spend five weeks working with Overton's unpublished data at the University of Lund, where Overton served as the first Professor of Pharmacology from 1907 until his retirement in 1930. Stephen Thesleff, who is the second person after Overton to hold the Chair of the Pharmacological Institute, alerted me to the existence of Overton's unpublished papers, obtained grant funds, made all the necessary arrangements for me to live and work in Lund, and served as my sponsor for this trip.

Since I have lived with Overton's work for several years now, it was especially meaningful during this time in Lund to have the opportunity to meet with his daughters, Harriet Overton and Margaret Overton-Haikola, who graciously shared their father's papers, books, and family photographs, as well as anecdotes about him. Professor Thesleff and the Overton family were all extremely hospitable and made Charles Ernest Overton come alive as a person as well as a scientist to me.

Overton indicated in his 1899 publication that he attempted to obtain samples of and test every chemical that was commercially available at that time, and the data published in this book represent only a portion of his studies. Plans are now being formulated for organizing these unpublished Overton data in Lund and making them available to the scientific community.

Overton's *Studien über die Narkose* is viewed by scientists in many different fields as the turning point in understanding the relationship between chemical structure and cell permeability, and between chemical structure and the potency of chemicals acting by an anaesthetic mechanism. (Hans Horst Meyer, working at the University of Marburg, independently and at the same time as Overton, also concluded that anaesthetic potency correlates with partition coefficient, and for this reason, this theory is commonly referred to as the Overton–Meyer or Meyer–Overton lipoid theory. (Lipnick, R.L. (1989) Hans Horst Meyer and the lipoid theory of narcosis, *Trends Pharmacol. Sci.*, **10**, 265–9.) I am grateful to the Wood Library-Museum of Anesthesiology for supporting the publication of this book, to Keith Miller, Peter Winter, and Leonard Firestone for contributing to this work, and to Drs Greene and Miller, and especially to my wife Anne R. Lipnick for proofreading the entire manuscript and for suggesting helpful changes.

Robert L. Lipnick
Alexandria, Virginia

Introduction

1

Introduction

Keith W. Miller

Overton's career was extraordinary in a number of ways. He took his PhD at Zürich in botany in 1889 and continued there as a lecturer in biology and assistant in botany until, in 1901, he moved to the Physiology Department at Würzburg. It was there, in 1907, that he accepted a Chair in Pharmacology at the University of Lund. How was it that he started out as a botanist and ended up as a pharmacologist – one whose work is still quoted in the three separate fields of physiology, pharmacology and toxicology?

Overton's seminal work was undertaken in the decade following the award of his PhD with his contribution to cell permeability studies culminating in a paper published in 1899 [1]. This paper made a number of important points, among them laying the foundation of what was to become the Meyer–Overton theory of narcosis. Here, however, I will emphasize his contribution to cellular permeability and the concept of the cell membrane. The 1899 paper [1] provided guidelines, referred to until this day as Overton's Rules, for predicting the relative cellular permeabilities of various substances, and it went on to note the simple relationship between lipid solubility and the ability of non-electrolytes to enter a cell. The latter relationship is illustrated with modern data in Fig. 1.1. While to our minds, conditioned by the lipid bilayer concept, this relationship is not too surprising, at the time it was published Overton's contribution must have seemed counter-intuitive.

The cell theory had been enunciated in 1838–9 by Schleiden and by Schwann [2]. One of the significant structural features of the cell was a membrane separating its contents from the surrounding fluids. However, the tools to unequivocally delve into cellular structure were not

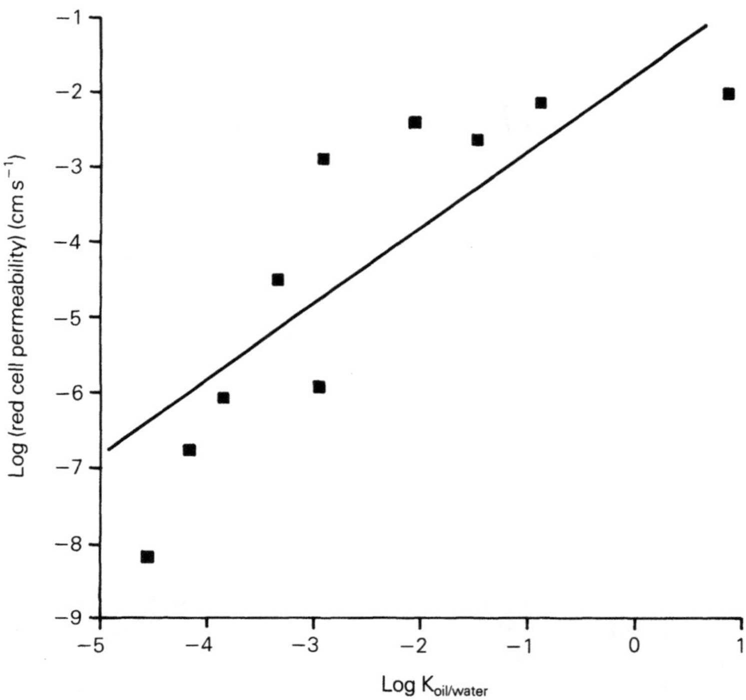

Fig. 1.1 Overton's Rule illustrated here by plotting on a double logarithmic scale data for the permeability of human red cells to ten non-electrolytes against their oil–water partition coefficients, $K_{oil/water}$. The linear regression through the data was fitted with a slope of 1.0, as required by Overton's Rule. From left to right the non-electrolytes are: erythritol, glycerol, urea, ethanediol, thiourea, water, methanol, ethanol, *n*-propanol, *n*-hexanol. (After Jain, M. (1988) *Introduction to Biological Membranes*, Wiley, New York, p. 121).

available until well into the twentieth century, and, in the meantime, biologists, armed with the light microscope and various dyes, came to consider the cell as a mass of protoplasm containing a nucleus. By the last decade of the nineteenth century, Verworn could state that the concept of a cell membrane had disappeared.

Against this tide, it was the osmotic properties of cells that brought the concept of the plasma membrane back into biology, and, of course, it was the plant physiologists and botanists who were most concerned with osmosis. It had been known since the middle of the century that cells of all origins shrunk when surrounded by solutions that were more concentrated than those in the interior of the cell (a process then referred to as plasmolysis). This suggested that the cell surface possessed a differential permeability to solvent and solute; the cell's solvent alone passing through the membrane to equalize the concentration on either side. The analogy with the semipermeable membrane of osmosis was clear and it was Pfeffer, who, considering the properties of plant cells in his classic book on osmotic pressure, first coined the term plasma membrane to describe a barrier he inferred to exist, even though he could not demonstrate it morphologically – a serious failing in the eyes of experimental biologists. In any event, the ideas of a cell membrane with holes of sufficient size to allow the passage of water could be said to exist well before Overton's time.

Another conceptually important piece of work preceding Overton's contribution was that of the German physicist, Georg Quincke. He studied oil droplets interacting with water and the myelin figures formed by lipids in water. Considering plasmolysis, he suggested in 1888 that the protoplasm boundary consisted of an enveloping fluid membrane too thin to be perceived by the light microscope (<100 nm). Since fatty oils or liquid fats were the only substances he knew to possess the property of forming such films, he concluded they must be the constituents of the plasma surface.

Such physical studies drew attacks from biologists, and, although Quincke defended himself stoutly against the 'descriptive' sciences, his ideas were apparently given little weight. Nonetheless, by 1897 Pfeffer seems to have taken in Quincke's point on thickness, going so far as to speculate about a single or double molecular layer, whilst firmly rejecting any idea of a film of oil. The stage was set for Overton.

Further plasmolysis studies had demonstrated that the concept that the plasma membrane (as we would now call it) behaved as a semipermeable membrane was incomplete. Thus, with certain solutes the shrunken cells gradually regained their original volume, suggesting that the plasma membrane was slowly permeable to such solutes. Overton [1] studied the rate of reversal of plasmolysis induced by hundreds of such solutes in plant cells, coming to the conclusion that '... the peculiar osmotic properties of living protoplasts are dependent on the phenomenon of "selective solubility" ...'. Moreover, he concluded that, 'the general osmotic properties of the cell are due to an impregnation of the boundary layers of the protoplast by a substance whose solvent properties for various compounds correspond overall to those of a fatty oil.' Since the lipid impregnated the 'boundary layers', the concept of a semipermeable membrane was retained while the lipid soluble solutes were able to diffuse through the non-porous lipid. Although Overton went on to speculate that it might be cholesterol and lecithin that impregnated the boundary layer, he was still a long way from the concept of a lipid bilayer which was to emerge much later after the coming of Langmuir's trough and Bragg's X-ray diffraction studies of fatty acids. Indeed, by the mid-1920s, although the concept of a lipid bilayer is clearly evident, there was no bold statement of a membrane model incorporating it. Kepner [2] has written

> ... suppose someone had a lipid model ready to go at this time, but paused first to consider who would be lying in ambush to savage it. On the one side, the colloid chemists would inquire into one's acquaintance with the properties of proteins. On the other side, many of the researchers studying permeability would shoot holes in the model until one acknowledged the need for pores. Additionally, the surface scientists would be interested in the reasons underlying one's particular choice for the arrangement of the lipid molecules, since several possibilities existed.

Two years after Overton's death, Danielli and Davson published their model with a clearly recognizable lipid bilayer. Even so, it was not until the mid-1960s, with the coming of Bangham's liposomes and various spectroscopic techniques, that the concept that a fluid lipid layer is a fundamental component of plasma membranes was finally accepted.

In this context it is, perhaps, not surprising that Overton's professional recognition came from the field of pharmacology rather than physiology, and that he was rather cautious when speculating on the nature of the structure of the plasma membrane. He states in this monograph (p. 52), 'Although individual theories on this subject differ rather widely, the living protoplasm can, *with little objection* be compared to a completely soaked sponge.' When he is on experimental ground, however, there is no hesitation. He succinctly describes (p. 51) the preparation of what we would now call multilamellar vesicles. 'Furthermore, a lecithin and cholesterol mixture formed by evaporation of a benzene solution, and then mixed with water, swells up ... and the cholesterol remains (if not present in too high a proportion) dissolved in the swollen lecithin.'

It is clear that with the tools at hand Overton could never have arrived at a provable molecular model of the plasma membrane. His great achievement was closer to that of the thermodynamicists of his era. He laid down two great rules, the Overton permeability rule and the Meyer–Overton rule of general anaesthesia. Each provides a first approximation that successfully accounts for properties that vary over some six orders of magnitude, leaving posterity to argue over the fine details. Read this monograph and be humble!

REFERENCES

1. Overton, E. (1899) Über die allgemeinen osmotischen Eigenschaften der Zelle ihre vermutlichen Ursachen und ihre Bedeutung für die Physiologie. *Vierteljahrsschr. Naturforsch. Ges. Zürich,* **44**, 88–135. Quoted in English from 'On the general osmotic properties of the cell, their probable origin and their significance for physiology', in *Cell Membrane Permeability and Transport* (ed. G.R. Kepner), Dowdon, Hutchinson and Ross Inc., Stroudsburg, PA, pp. 29–56.
2. For a review of this and other work mentioned see Kepner, G.R. (1983) From oil layer to bilayer in *Liposome Letters* (ed. A.D. Bangham), Academic Press, London, pp. 15–27 and references cited therein.

2
Introduction

Peter M. Winter and
Leonard L. Firestone

It is unnecessary to describe the horror of surgery prior to the demonstration of ether anaesthesia at Massachusetts General Hospital in 1846. It was clearly a nightmare for patients and must have been little better for the medical personnel involved in so terrifying an undertaking. What is less obvious is that the introduction of anaesthesia accomplished far more than the abolition of intraoperative pain. The solution of the problem of pain enabled the evolution of virtually all of modern surgical therapeutics. Prior to this development, the major characteristic of a technically brilliant surgeon was speed – the ability to do a below-the-knee amputation in less than a minute made or unmade reputations. Intra-cavitary surgery; operations in the chest, abdomen or skull were largly unthinkable and when attempted, commonly led to the death of the patient, not because of pain *per se*, but because the surgeon had no time in which to think and take deliberate action.

So practical were the properties of anaesthetics, that their clinical use spread rapidly throughout the medical world without the least understanding of the mechanism(s) by which the agents worked. The drugs obviously produced unconsciousness and freedom from the perception of noxious stimuli. It was also desirable that they did so as rapidly as possible, and that such effects were completely reversible with few physiological side effects.

We would not dispute these requirements today. Within the context of then-current chemical knowledge, three agents appeared to fit all or some of this description. Diethyl ether, as used by Morton, became the standard for generations. Nitrous oxide provided all the correct attributes but one – sufficient potency to cause unconsciousness and surgical anaesthesia. Chloroform also provided the requisite analgesia and unconsciousness and was used for decades, despite its potentially lethal side effects.

The early lack of understanding and, indeed, concern about anaesthetic mechanisms of action should not be too surprising. Very few of the drugs then in use were understood in any detail. Drugs were found largely by trial and error in animals and humans. That they worked and were relatively safe was all that was required.

In the context of this pragmatic medical world, Ernest Overton was a fascinating exception. Born in Cheshire, England, in 1865, Overton was a distant relative of Charles Darwin. His maternal grandfather, Reverend W. Darwin Fox was an entomologist, second cousin and close friend of Darwin. With his family, Overton moved to Switzerland at the age of seventeen and there completed his education. He received a PhD in Botany in 1899 from the University of Zurich having worked primarily with Professor Arnold Dodel. In 1907 he accepted the Chair in Pharmacology at the University of Lund, where he was to stay for the rest of his career.

Ernest Overton is primarily known in modern anaesthesiology for what has been called the Meyer–Overton rule of anaesthesia (Fig. 2.1), work which he published in 1899 and which was independently developed and published by Professor H.H. Meyer. This remarkably durable but mechanistically unenlightening correlation relates the potency of an anaesthetic to its solubility in lipid. The result gives an astonishingly close correlation over a wide potency range (three orders of magnitude for the data illustrated). The most potent inhalation anaesthetic known today, thiomethoxyflurane, has an oil/gas partition coefficient of 7230 and a MAC (minimum alveolar concentration, a measure of anaesthetic potency) of 0.035% atmospheres. The relatively impotent anaesthetic N_2O has a MAC of 1.05 atmospheres. Even noble gases not usually thought of as anaesthetics, such as nitrogen and hydrogen, have an effect predicted by their lipid solubility. Being relatively lipid insoluble, many atmospheres of these gases are required to produce obtundation. Thus, the lipid solubility-potency prediction for these gases must be corrected for the 'anti-anaesthetic' effect of increased pressure [1]. Once this is taken into account, however, the predictions become accurate. The anaesthetic potency for nitrogen has been measured to be about 40 atmospheres in mammals [2]. Sub-anaesthetic concentrations are the cause of the 'nitrogen narcosis' experienced by divers breathing compressed air. The anaesthetic effect of nitrogen is the depth-limiting factor in the use of air as a diving breathing mixture. At greater depths,

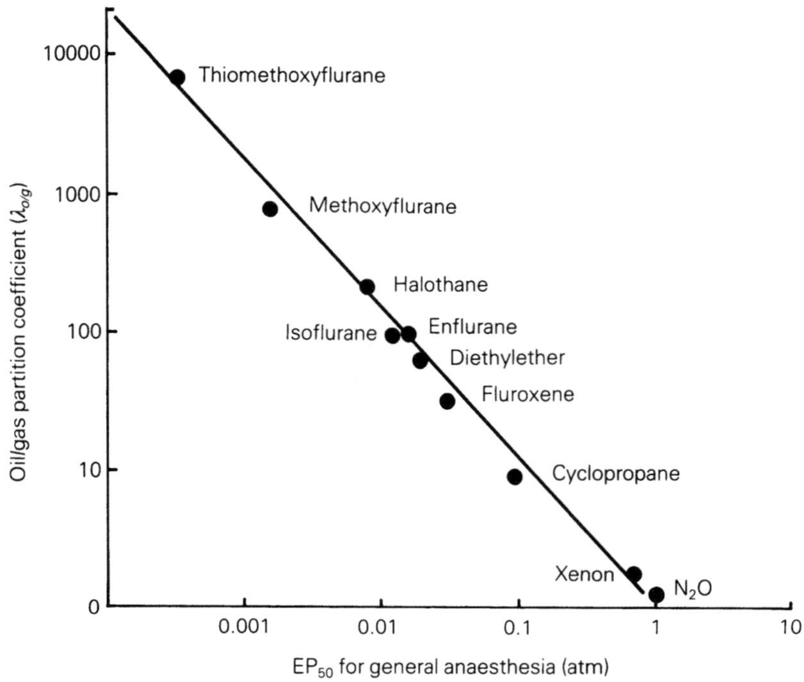

Fig. 2.1 The Meyer–Overton rule of anaesthesia. A logarithmic plot of EP_{50} versus $\lambda_{o/g}$ is represented for structurally heterogeneous general anaesthetics. (EP_{50} is the median effective partial pressure in man which inhibits movement in response to a surgical stimulus and $\lambda_{o/g}$ is the oil/gas partition coefficient). A free, least squares fit of the data reveals a close correlation between the variables: the straight line has a slope of -1.07 ± 0.04 and a correlation coefficient (r) of -0.996. The EP_{50} for thiomethoxyflurane, a toxic volatile anaesthetic not in clinical use, was determined in dogs. The oil/gas partition coefficient ($\lambda_{o/g}$) is: litres of gas (expressed at 1 atm and 37°C) dissolved in 1 litre of solvent when the partial pressure of the gas above it is 1 atm. Determinations were in either olive oil or octanol, whose drug absorption properties are similar to those of the lipids in biological membranes. Sources for all values are compiled in reference 2.

helium and oxygen mixtures are used because the lipid solubility of helium is so low that it does not cause narcosis at any realistic depth. Based on the same calculation, even the nitrogen in air at atmospheric pressure may be shown to possess a depressant effect [3].

Overton's original interest was not in anaesthetics *per se*, but rather in cell permeability. Stemming from an interest in mechanisms of heredity, he was concerned with substances capable of rapid penetration of living cells. It must be remembered that in the biology of Overton's times, little was known of cell membranes or of their composition. Yet they seemed to allow living cells to resolve an apparently conflicting set of requirements, differentiating self from environment, while at the same time, permitting rapid, simultaneous exchange of nutrients and wastes with the environment.

Indeed, it can be said that Overton's interest in anaesthetics originated when he determined that these agents could be used as probes, or markers, in his studies of cellular physiology: 'Narcosis phenomena are not only of enormous interest to the student of pharmacology but are also fundamental to biology in general and especially to cell physiology' (p. 26). This tradition continued into modern times with Ferguson's investigations of the bacterial actions of anaesthetic alcohols [4], and still flourishes today in the multidisciplinary neurosciences.

Although Overton was not the first to concern himself with and study the mechanisms of anaesthetic action, he was probably the first to quantify and compare their potencies. As stated in the *Studien* '... it is much more practical to use as a measure of the relative narcotic strength the lowest partial pressure of the various anaesthetics that are sufficient to produce narcosis' (p. 46). He asserts further that this measure is more important than the ratios of the amounts of anaesthetics that must be mixed with air in order to achieve narcosis: 'Of primary importance is the concentration of the anaesthetic in the blood plasma, in the intercellular lymph and in the fluid surrounding the ganglia cells' (p. 46). Overton was also an able chemist and understood lucidly the concept of distribution coefficients and described their determination. And, he was among the first to use quantitative dose-response curves in biology.

In this monograph, Overton elegantly reviews the state, at the end of the nineteenth century, of thought about anaesthetic mechanisms. Claude Bernard, one of the most original and insightful scientists of his time, thought that anaesthetics acted by partial and reversible coagulation of cell protoplasm. Overton also refers to the hypothesis generated in 1847 by Bibra and Harless, who described the abilities of various anaesthetics to dissolve fats. They believed that anaesthetics worked by dissolving fat out of the brain and increasing it in the liver. Overton politely calls this 'a chance occurrence in the analytical results' (p. 68) and asserts that the reversibility of anaesthetics proves that the hypothesis cannot be correct.

He goes on to say, with astonishing insight, 'It is also very probable that the non-specific narcotics exert their effect principally on the cholesterol and lecithin-related constituents of the cells, but not in the way visualized by Bibra and Harless. These compounds probably change the normal physical state of these cell constituents without causing them to be removed from the cells' (p. 70).

As readers of his work will learn, Overton, in the pursuit of his hypothesis of anaesthetic mechanisms, anticipated many clinical concepts later described more formally. He understood that opiates and inhalation anaesthetics were different classes of drugs, and referred to the former as 'basic narcotics' and the latter as 'non-specific narcotics'. Overton's idea to employ a combination of opiates and gaseous anaesthetics to induce clinical anaesthesia represents a technique that is now routinely applied. This procedure is referred to as 'balanced anaesthesia' by contemporary clinical anaesthesiologists. He discussed the additivity of inhalation anaesthetics when considering drugs with a lethal dose below their anaesthetic dose, using additivity with drugs of known potency to derive the anaesthetic dose of the toxic substance.

He also understood something of the uptake and distribution of inhalation agents. Part of the appeal of using inhalation anaesthetics when studying drug effects, is the ability, after equilibration, to maintain a constant brain concentration of the agent, independent of metabolism and redistribution. Overton was aware of the importance of such control, and recognized that it could be achieved in aquatic animals by dissolving known concentrations of an agent in the aquarium water. Because of this, he found tadpoles to be useful experimental animals, and, indeed, his reasoning is still embraced by current investigators.

In his talk to The Leopoldina commemorating the one hundredth anniversary of the birth of Overton, Paul Runar Collander said, 'Like Darwin, Overton had a striking intuitive ability to recognize the great fundamental problems and to see how they should be solved [5]. With training in botany, not medicine, he was freed of the pragmatic concerns of practitioners and pursued interests which have given us unique insights into one of the great riddles in medical science. Overton's work elegantly pointed subsequent investigation to a lipophilic site of action, and was such a conceptual leap, it has been asserted that little more progress has been made since his death in 1933. Indeed, the unyielding nature of this problem serves to strengthen our appreciation of the 'careful logic' [4] of the remarkable early observations contained in this volume. And for facilitating this appreciation, the anaesthesia world owes much to Dr Lipnick.

REFERENCES

1. Dodson, B.A., Furmaniak, Z.W. and Miller, K.W. (1985) The physiological effects of hydrostatic pressure are not equivalent to those of helium pressure on *Rana pipiens*. *J. Physiol.*, **362**, 233–44.

2. Firestone, L.L., Miller, J.C. and Miller, K.W. (1986) Tables of physical and pharmacological properties of anesthetics, in *Molecular and Cellular Mechanisms of Anesthetics*. (eds S.H. Roth and K.W. Miller), Plenum, New York, pp. 455–70 and references cited therein.
3. Winter, P.M., Bruce, D.L., Bach, M.J. *et al*. (1975) The anesthetic effect of air at atmospheric pressure, *Anesthesiology*, **42**, 658–61.
4. Ferguson, J. (1939) *Proc. Roy. Soc. B.*, **127**, 389.
5. Collander, P.R. (1962–3) *Leopoldina*, 3rd series, pp. 8–9, 242–54.

3

Charles Ernest Overton: narcosis studies and a contribution to general pharmacology

Robert L. Lipnick

The year 1986 marked the 85th anniversary of the publication of Charles Ernest Overton's classic monograph *Studien über die Narkose* ('Studies of Narcosis') [1]. This book has been cited widely by scientists studying the correlation of biological activity with partition coefficients and the mechanism of anaesthesia. It is of enormous value to modern toxicologists, particularly those involved in the development of quantitative structure–toxicity relationships, and the predictive limitations of such models. Overton reported a wealth of precise physiological and histological observations on the response of a variety of plant and animal species to a large number of organic compounds. His remarkable deductions continue to serve as an inspiration for new scientific research and discovery.

Overton was born in England in 1865 and moved to Switzerland in 1882, where he received a PhD in botany at the University of Zürich in 1889. In 1907, he accepted the Chair of Pharmacology at the University of Lund in Sweden, where he remained until his death in 1933 [2, 3]. Overton was a brilliant and versatile researcher whose work in narcosis and other areas presaged important discoveries in many areas of modern pharmacology. Most of his publications are in German and, despite the

Fig. 3.1 Charles Ernest Overton (from [3]).

fact that only a few excerpts of the narcosis book have, until now, been available in English translation [4], it has been cited frequently over the years and up to the present by numerous authors in a variety of fields including quantitative structure–activity relationships, pharmacology, anaesthesiology, medicine, toxicology and biophysics. Working alone, Overton performed most of the experiments reported in his book between 1890 and 1898; they evolved from systematic studies published in 1895 and 1896 of the permeability of living plant and animal cells to a large number of organic compounds [5, 6]. He first presented his theory of narcosis in a lecture to the Society for Natural History Research in Zürich on October 31, 1898, and published this work in 1899 [7], the same year that Hans Horst Meyer [8] and his collaborator Fritz Baum [9] independently set forth what was essentially the same hypothesis.

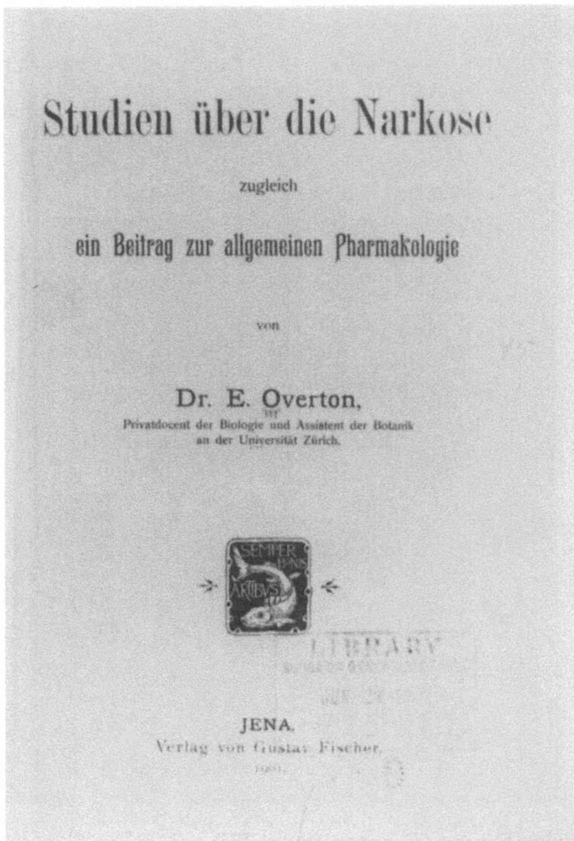
Studien über die Narkose

zugleich

ein Beitrag zur allgemeinen Pharmakologie

von

Dr. E. Overton,

Privatdocent der Biologie und Assistent der Botanik
an der Universität Zürich.

JENA.
Verlag von Gustav Fischer.

Fig. 3.2 Title page of *Studien über die Narkose.* (Courtesy of the National Library of Medicine.)

3.1 NARCOSIS THEORY

Overton asserted that a viable theory of narcosis must account for the complete and rapid reversibility of narcosis following removal of a narcotic or anaesthetic agent. As the starting point for his new theory, he chose the solubility of narcotics in cholesterol, lecithin and other lipoid substances contained in cells. Since the narcotics produce no observable chemical change upon these fatty substances, he attributed narcosis solely to physical changes induced from solution of the narcotic in these lipoid constituents. Although this explanation would predict that the intensity of effect is related primarily to the quantity of narcotic absorbed,

Overton did not rule out the possibility that the qualitative nature of the compound itself could play some role. He postulated that the intensity could be proportional to either the number of narcotic molecules absorbed or the volume which they occupy at the site of action. The latter hypothesis may have served as the inspiration to Mullins in investigating the thermodynamic consequences of a volume fraction theory of narcosis [10]. Overton also suggested that water molecules may be displaced from the lipoid phase following absorption, in a fashion not explained simply by solution theory. The clathrate theory of narcosis proposed independently by Pauling [11] and Miller [12] can be viewed as an expansion of Overton's early speculation.

3.2 ANAESTHETICS AND NARCOTICS

Overton considered the frequent distinction made by his contemporaries between anaesthetics and narcotics to be both artificial and inconsistent. Anaesthetics were considered to affect all types of plant and animal cells in a reversible fashion, while narcotics were believed to act only upon animal ganglia cells, and not necessarily in a reversible fashion. Overton found that the concentration required to suppress activity, that is, produce narcosis in plant cells, was generally six to ten times greater than that required to anaesthetize brain cells in higher animals. Therefore, those substances producing narcosis in animals at almost saturated solution are limited by their water solubility with respect to their ability to narcotize plant cells. He observed that substances of such limited solubility may or may not produce narcosis in cold blooded animals depending upon the temperature at which the experiment is conducted. Furthermore, substances of low solubility, producing no narcotic effect on either plant or animal cells at saturation, were found to reduce the concentration required of a second narcotic through their additive contribution.

3.3 BASIC ORGANIC COMPOUNDS

Overton found that in contrast to the non-electrolyte narcotics, basic organic compounds including certain alkaloids produce strikingly different effects which vary greatly both qualitatively and quantitatively depending upon the test organism. He attributed such variations to each compound's ability to form salt-like complexes with proteins. He considered differences in toxicity between such basic compounds to reflect variations in the solubility properties of their corresponding protein complexes, which may be similar for two alkaloids of different chemical constitution. He accounted for variations in organism sensitivity by the small differences in the chemical structure of proteins that correspond histologically and physiologically to one another. Thus, while a 1 : 2000 saturated aqueous solution of morphine produces almost no effect upon tadpoles, the maximum tolerated dose in humans of 0.1 gram would correspond to about 1 : 400 000, if distributed throughout the body.

3.4 ROUTE OF ADMINISTRATION

Overton made no distinction between anaesthesia produced via inhalation and narcosis produced by other routes of administration, since he thought that in each case, a chemical acts by a common mechanism and reaches the same site of action, i.e. the brain cells, from the blood in the same fashion. He considered the pathways by which narcosis-inducing substances, i.e. narcotics, enter the bloodstream to be unimportant so long as they achieve the required concentration in the blood.

Overton interpreted the variation in toxic response observed with administration via the stomach, rectum, skin or peritoneum to be a manifestation of different rates of absorption. To calculate an approximate concentration in the bloodstream, Overton concluded that knowledge would be required of the ability of a substance to penetrate various tissue cells, the approximate fat content of the animal, partition coefficients between water and the fats in question, metabolism, excretion and volatilization via the lung.

3.5 EXPERIMENTS AT KNOWN AND CONSTANT BLOOD PLASMA CONCENTRATION

Overton found that he could study toxicity at a known blood plasma concentration if compounds were administered via the lungs at a constant and known partial pressure and temperature. Similarly, the blood plasma or cellular fluid concentration could be defined in corresponding controlled experiments conducted on aquatic plants and animals submerged in a solution of known and fixed concentration and temperature, and of sufficient volume to ensure that the concentration would not change appreciably as a result of uptake during the experiment. In general, he found that the time required to produce an effect was similar for both gaseous and aquatic routes of administration.

Overton determined that an ether concentration of $20\,g\,hl^{-1}$ in air was just sufficient to maintain complete narcosis in dogs, which have a body temperature of approximately 38°C. Based upon the Henry–Dalton Law, he calculated that at equilibrium this vapour concentration would produce a 0.25% concentration of ether in the blood plasma, the same concentration he found sufficient to produce narcosis in tadpoles.

3.6 TEST ORGANISMS AND CHEMICALS

Overton reported test results for over 130 compounds in this book. However, this represents only a portion of the data from which his conclusions are drawn for he had attempted to obtain samples of and test every organic compound that was commercially available at the time. Overton employed algae and a wide variety of aquatic animals including tadpoles, daphnia, fish, crustaceans, bryozoa, and annelids to study toxicity at a constant blood plasma concentration. Most of the experiments which he reported in detail were conducted using tadpoles of the species *Rana temporaria*. The compounds tested included monohydric, dihydric, and polyhydric alcohols, aliphatic and aromatic hydrocarbons, nitriles, nitroparaffins, aldehydes, ketones, sulfones, esters of organic and mineral acids, various aromatic compounds, amines and alkaloids.

3.7 NARCOTICS AND ORGANIC BASES

Overton found that most of the organic non-electrolytes that he tested produced narcosis. By contrast, organic bases such as alkaloids generally did not produce narcotic effects. He observed no sharp delineation, but rather a continuum of effects between the neutral and basic classes of organic compounds. For example, ethyl alcohol exhibited almost purely narcotic effects. Tadpoles could be kept in a 1% solution of ethyl alcohol for days without showing any narcosis or ill effect. However, in a 1.5% solution, they became narcotized in 2–3 minutes and could be maintained in this state for 20 hours; the narcosis faded once they were returned to pure water.

3.8 ESTER HYDROLYSIS

Overton found that esters of monovalent aliphatic acids behaved in an interesting fashion with respect to the length of time a tadpole could be maintained in a narcotic state without dying. For example, he reported that the length of undisturbed narcosis for amyl acetate was 2.5–3 hours, but for ethyl valerate, which has the same chainlength, it was 15 hours. Within each homologous series of esters (e.g. acetates), he found this duration to be a reflection of the rate of metabolic ester cleavage within the tadpole. Overton confirmed and expanded these findings for a more diverse group of esters in a later study [13].

3.9 PROGRESSIVE TOXICITY

Overton observed that some compounds such as hydrocyanic acid and the lower molecular weight monovalent aldehydes produced an initial rapid response resembling narcosis, but required a much longer period of time to produce their maximum effect. He ascribed the second response to a slow progressive chemical reaction with one or more cellular constituents.

3.10 CHEMICAL STRUCTURE AND NARCOTIC POTENCY

Overton discovered that substances having little or no solubility in a mixture of cholesterol and lecithin, such as alcohols containing four or more hydroxyl groups, produced no narcotic response in tadpoles at any concentration. He concluded that no equalization of the concentration in the blood and the external aqueous phase could be achieved for such substances as long as the animal was alive. He discovered a systematic increase in narcotic potency with increasing chainlength among groups of related compounds. Beyond a certain chainlength, however, he noted that narcotic properties could no longer be detected. Within a series of isomers, he observed that the isomer with the least branching or which was furthest removed from a spherical shape had the strongest activity. In addition, replacement of a hydrogen or halide by hydroxyl reduced activity. Similarly, he noted that in general, activity increased in the order iodide > bromide > chloride.

3.11 PARTITION COEFFICIENT AND NARCOTIC POTENCY

From these experiments, Overton concluded that only one physical property, the lipoid–water partition coefficient, changed in a way that reflected the regularities he observed between chemical structure and narcosis action. Linear correlations between *n*-octanol–water partition coefficient and molar toxicity on a log log scale subsequently derived for a variety of organisms have confirmed Overton's findings [14]. A study using Overton's original data provided a high correlation ($r^2 = 0.913$) in this type of analysis, but a much poorer correlation with other parameters (polarizability, molar attraction, parachor and molecular weight) [15].

3.12 CHEMICAL STRUCTURE AND PARTITION COEFFICIENT

Overton's early observations regarding a systematic trend in the contribution of structure fragments to partition coefficient have now been confirmed and amplified as the fragment constant methodology [16, 17]. The recent computerization of this fragment constant methodology [18] has given a strong impetus to the advancement of predictive toxicology and pharmacology [19, 20].

3.13 LIMITING SOLUBILITY

Overton found that within a homologous series, although the partition coefficient continues to increase with chainlength, the absolute solubility in oil or a mixture of cholesterol and lecithin at room or blood temperatures decreases rapidly beyond a certain point in the series. For example, phenanthrene, which is readily soluble in olive oil and related substances at room temperature, is a narcotic, but anthracene, an isomer, is not soluble and does not show narcotic effects. Overton concluded that low water solubility alone will not limit narcotic toxicity, as in the case of phenanthrene which dissolves in about 300 000–400 000 parts of water, but produces narcosis at one part in 1 500 000. For experiments conducted at this very low concentration, 36 hours were required for complete narcosis to take place, which Overton accounted for based upon the slow rate of transport and accumulation of phenanthrene into the ganglia cells.

Overton concluded from his physiological experiments, and a large number of measurements of partition coefficients and solubilities in water and oil phases, that the narcotic strength depends primarily upon its partition coefficient between water and the lipoid substances in cells.

Overton's ingenious experiments and remarkable deductions have provided a stepping stone for work by investigators in many disciplines, and are expected to continue serving as an inspiration for new scientific research and discovery. His book is of enormous value to modern toxicologists, particularly those involved in the development of quantitative structure–toxicity relationships and the limitations of such models.

Acknowledgement

The author is grateful to Dr James H. Gilford, Chief, Environmental Effects Branch, and Dr Irwin Baumel, Director, Health and Environmental Review Division, EPA; and to Professor Roelof R. Rekker, Department of Pharmacochemistry, Vrije Universiteit, Amsterdam, for their encouragement in the preparation of this article.

REFERENCES

1. Overton, E. (1901) *Studien über die Narkose, zugleich ein Beitrag zur allgemeiner Pharmakologie*, Gustav Fischer, Jena.

2. Collander, P.R. (1962–3) *Leopoldina* 3rd ser., 8–9, 242–54.
3. Thunberg, T. (1933) *Kungl. Fysiografiskla Sällskapets i Lund Förhandlingar*, **3**, 45–52.
4. Holmstedt, B. and Liljestrand (1981) *Readings in Pharmacology*, Raven Press, New York, pp. 150–4.
5. Overton, E. (1895) *Vierteljahrsschr. Naturforsch. Ges. Zürich*, **40**, 159–201.
6. Overton, E. (1896) *Vierteljahrsschr. Naturforsch. Ges. Zürich*, **41**, 383–406.
5. Overton, E. (1899) *Vierteljahrsschr. Naturforsch. Ges. Zürich*, **44**, 88–135.
8. Meyer, H. (1899) *Arch. Exp. Pathol. Pharmakol. (Naunyn-Schmied.)*, **42**, 109–18.
9. Baum, F. (1899) *Arch. Exp. Pathol. Pharmakol. (Naunyn-Schmied.)*, **42**, 119–37.
10. Mullins, L.J. (1954) *Chem. Rev.*, **54**, 289–323.
11. Pauling, L. (1961) *Science*, **134**, 15–21.
12. Miller, S.L. (1961) *Proc. Nat. Acad. Sci., USA*, **47**, 1515–24.
13. Overton, E. (1925) *Scand. Arch. Physiol.*, **46**, 333–4.
14. Hansch, C. and Dunn, W.J., III (1972) *J. Pharm. Sci.*, **61**, 1–19.
15. Leo, A., Hansch, C. and Church, C. (1969) *J. Med. Chem.*, **12**, 766–71.
16. Hansch, C. and Leo, A. (1979) *Substituent Constants for Correlation Analysis in Chemistry and Biology*, Wiley-Interscience, New York.
17. Rekker, R.F. (1977) *The Hydrophobic Fragmental Constant*, Elsevier, Amsterdam.
18. Leo, A. and Weininger, D. (1985) *Medchem Software Release 3.3*, Medicinal Chemistry Project, Pomona College, Claremont, California.
19. Lipnick, R.L. (1985) in *QSAR in Toxicology and Xenobiochemistry* (ed. M. Tichy), Elsevier, Amsterdam, pp. 39–52.
20. Lipnick, R.L., Pritzker, C.S. and Bentley, D.L. (1985) in *QSAR and Strategies in the Design of Bioactive Compounds* (ed. J.K. Seydel), VCH, Weinheim, pp. 420–3.

Studies of Narcosis and a Contribution to General Pharmacology

Charles Ernest Overton

Preface

My studies of the osmotic properties of living plant and animal cells have been in progress for many years and have covered almost all commercially available organic compounds. During these studies, I have had the opportunity to make a large number of observations of narcosis in both plants and animals. These observations and the experiments that resulted from them have made it possible to determine, for the non-specifically acting compounds, the properties of each compound that characterize it as a narcotic. These studies have also made it possible to predict, at least approximately, the relative strengths of the individual narcotics from certain physical properties.

I intended originally, after finishing my studies of osmosis, to undertake a larger comparative study of the effects of narcotics, antipyretics and antiseptics and the mechanism of these effects, and then to set down the results of my studies of narcotics in that work as well. In spite of the wealth of material that had already been gathered for such a work through my studies of osmosis, more time was required before it could be published. Therefore, I decided meanwhile to publish in shorter form the more significant findings made thus far; namely, those related to the mechanism of narcosis. I had already begun work on editing when, through the kindness of Professor M.V. Frey, who was already familiar with my views on the mechanism of narcosis, I became aware of a short report by Professor Hans Meyer in Marburg. Approaching the subject from another angle and working with other material relevant to the topic, Professor Meyer had reached substantially the same view as mine on the production of narcosis by non-specific narcotics. Through the great kindness of Professor Meyer, I received the proofs of more detailed reports on this subject before his studies appeared in print and I would like to take this opportunity to express my sincere thanks to him.

Various circumstances have combined to delay the publication of this work. First, the editing was interrupted in order to await the results of certain tests I was conducting with lecithin and cholesterol during the first half of last year (1899). These experiments are very pertinent to the subject of this book. When these experiments were finished, I had hoped to complete the editing of this work, but was again prevented due to a long illness.

I developed my theory on narcosis entirely independently of the prevailing hypotheses on narcotics, which were unknown to me at the time. Nevertheless, it seemed appropriate both to provide an overview and to discuss critically these earlier hypotheses.

Almost all of my own experiments were carried out between 1890 and 1898 and were initiated simply to study osmosis and not to investigate the mechanism of narcosis. This latter subject was brought to my attention in the course of my studies and experimental results related to my theory of the osmotic properties of the cell. During the editing of this work, I repeated some of my earlier experiments. These new experiments provided results that corresponded completely with the earlier ones. The only experiments not performed until this year (1900) were those on the partial pressure of ether and chloroform in an air mixture that is necessary to produce narcosis at various temperatures in cold-blooded animals. For years I have applied the principles underlying these experiments for other purposes.

Narcosis phenomena are not only of enormous interest to the student of pharmacology but are also fundamental to biology in general and especially to cell physiology. Their significance has been gaining increasing recognition. In this work, narcosis is, in fact, treated mainly from this more general point of view. It therefore seemed appropriate to make their findings available in the form of a monograph, which has been possible with the willing co-operation of the publisher.

I am greatly indebted to Professor Dodel for his support in reading a portion of the proofs.

Finally, it should be noted that a small part of the data on narcotics included in this work, after receiving written permission from the author, has already been published by Dr E. Rost, partly in his collected reference work on narcosis for *Friedlanders Fortschritte der Medicin* and partly in the *Naturwissenschaftlichen Rundschau* (vol. XIV, 1899).

Charles Ernest Overton
Zürich, December 1900

Part One
General Section

1
Background

1.1 INTRODUCTION

Since antiquity, the power of certain organic substances contained in the milky juices of some plants to relieve pain and induce sleep has captured the interest and imagination of doctors and lay people alike. This interest is reflected in the tales of poets from many countries. I will remind you here only of the occasion in the *Odyssey* when Helen mixes a remedy in the wine of the grieving Telemachus '... to banish sorrow and hatred and the memory of all pain'; various tales of Boccaccio in the *Decameron*, e.g. the twenty-eighth story in which an abbot administered a powder to Ferondo that was said to make him sleep for three days and appear as if dead; also, the fortieth tale in the same collection; and finally Act IV, Scene I of Shakespeare's *Romeo and Juliet*.

When considering whether these literary references constitute historical proof that narcosis was used in earlier times, one should not overlook the fact that the poets on these occasions furnished their characters with much greater knowledge and more remedies than they could possibly have actually possessed according to the general state of affairs at the time. Usually, it was the actual and legendary effects of opium and hashish (a preparation made from *Cannabis indica*) that were such a rich source of inspiration to the author's imagination.

There can be no doubt that since ancient times there have been cases in which, either by accidental or intentional poisoning or as a result of a disease of the nervous system, people have lapsed for longer or shorter periods of time into a state in which they were insensitive to anything happening to them. It is no less certain that until the 1840s doctors knew of no sure, generally applicable, method of inducing this state at will. Up to this time, the most scientifically knowledgeable surgeons had emphatically denied the possibility of operating without pain or seriously risking the life of the patient and had refused to try the various methods proposed for producing insensitivity. In fact, Humphrey Davy had already shown that nitrous oxide, in addition to its other properties, also appeared to suppress pain. Thus, this compound, not being a harmful one, might be suitable for use in certain surgical operations. It was left, however, to the dentist, Horace Wells, to translate this suggestion into practice. Inspired by Wells' experiments, Morton and Jackson then tried to find other compounds that would produce a more complete and longer lasting narcosis than nitrous oxide. It is well known that in 1846 they found in ethyl ether a compound with the desired properties.

This date marks the beginning of scientific research into narcotics and narcosis, research that has continued to the present with uninterrupted zeal and has brought to light abundant material. These investigations led to the discovery of numerous compounds with narcotizing properties similar to those of ethyl ether and revealed the risks that are peculiar to each of the narcotics; they also have shown to a certain extent how to overcome these risks.

Thus far the designations narcotics and narcosis have been used in their broadest sense and they will continue to be used in this fashion in the remainder of this book. Meanwhile, it will be valuable to review what experiments have already been done in order to define the concept of narcosis, and in addition, to critically evaluate these experiments.

1.2 ATTEMPTS TO DISTINGUISH BETWEEN ANAESTHETICS AND NARCOTICS

Many authors, particularly the French* have attempted to distinguish clearly between anaesthetics and narcotics. Claude Bernard demonstrated that chloroform and ether suppress not only the functions of the ganglia cells but also, in sufficient concentrations, the action of muscle fibres, ciliary cells and other tissue constituents. Other substances, however, such as morphine and related compounds, paralyse the ganglia cells of the brain but have no effect, for example, on ciliary cells. Claude Bernard used only salts of morphine in his experiments and not the free base itself. We will return to this subject later.

Claude Bernard's findings were used by some of his followers as a basis for making a sharp distinction between anaesthetics and narcotics. Anaesthetics were considered to be distinguished first by their effect on all types of both animal and plant cells, and secondly by the reversibility of their effect. Narcotics, by contrast, were not considered to affect all protoplasts, and their site of action was attributed to the ganglia cells.

* According to Dastre [1], 'In principle, the action of narcotics is neither temporary nor universal.'

Furthermore, in contrast to anaesthetics, the effect of narcotics was considered, at least in principle, to not be temporary.

If, after extensive comparative experiments had been carried out with appropriate critical care, it had really turned out that two groups of compounds could be distinguished on a physiological basis, of which one group possessed the properties that were ascribed to narcotics and the other corresponded to anaesthetics according to the above definitions, then admittedly, such a definition between narcotics and anaesthetics would be justified.

Those investigators who proposed these definitions, however, have never carried out comprehensive research. The few experiments that have been performed in this field have proven to be lacking in both method and correct interpretation. This will be shown later in the discussion of the individual narcotics that follows in the experimental section of this book.

In fact, the above definitions of anaesthetics and narcotics were regarded even by their authors to a certain extent as theoretical definitions. Thus, for the actual classification of a given compound as either anaesthetic or narcotic, entirely different and usually arbitrary criteria have been employed.

Based upon the distinction made above between anaesthetics and narcotics, almost all the compounds classified in the narrow sense as narcotics would be termed anaesthetics. Those authors who wish to distinguish between narcotics and anaesthetics would classify the narcotizing compounds that are stable at room temperature, such as the urethanes, not as anaesthetics, but as narcotics. The entire class of urethanes, however, has an equally narcotizing effect upon plant cells and ciliary cells as on ganglia cells. If specific compounds that are classified as narcotics such as sulphonal, actually appear to have no noticeable narcotizing effect for example on plant cells and ciliary cells under normal experimental conditions, this is simply because the saturated aqueous solutions of the relevant compounds are not yet sufficiently concentrated to produce narcosis. For all typical anaesthetics, such as chloroform and ether, the concentration sufficient to suppress brain cell function in higher animals in much lower than the corresponding concentration needed for the complete narcotization of plant cells. Usually this ratio is between 1:6 and 1:10. It therefore seems obvious that those narcotics which need to be administered in almost saturated form in order to produce narcosis in the brain, cannot induce narcosis in plant cells and ciliary cells. It has probably not occurred to anyone to classify specific compounds within a special pharmacological group based upon their similar solubilities. If this was done, then in experiments with cold-blooded animals, the same compound would sometimes belong to one pharmacological class, and sometimes to another, according to whether the experiment was carried out at a higher or lower temperature. Added to which is the notable characteristic that in many cases, as a result of its low solubility, a compound by itself may have no observable narcotic effect on plant cells or on animal tissue cells of lower function. Neverthless, the compound may in fact exercise a narcotic effect on these cells if a second narcotic is added to it in solution. Complete narcosis occurs at lower concentrations of the second narcotic than would be required if the second narcotic were used alone, since the narcotic effects of both compounds are combined. This is a method that I have employed frequently in my studies of narcosis.

At least as far as the non-specific substances are concerned, i.e. compounds that are not of the nature either of a base, an acid or a salt, after carrying out numerous experiments, I have no hesitation in saying that up to now no compounds are known that would correspond to narcotics according to the definition of the French physiologists. Thus, all non-specific compounds that narcotize the brain cells also have a narcotizing effect on plant cells, ciliary cells, muscle fibres, and other parts of organisms, if the concentration of the compounds is sufficient. Furthermore, it is the case with all non-specific narcotics that their effect, if it was not too intensive, is temporary, i.e. it ceases after the relevant compounds have been completely removed from the fluid surrounding the cells.

Among those experiments that attempted to characterize anaesthetics according to their chemical and physical properties I will mention here only that of Raphael Dubois [2] which at least belongs among the more successful experiments in this direction. According to Dubois, anaesthetics are colourless and fragrant compounds with a stinging taste that produce a more or less burning sensation of warmth upon contact with the mucous membranes. They are volatile mobile liquids that in general are stronger anaesthetics the greater their vapour pressure and the lower their solubility. Their specific heat is low, usually much lower than that of water. Finally, they are dysosmotic, i.e. they penetrate organic membranes only with difficulty.

Various objections can be made to this characterization. The liquid anaesthetics in their pure undissolved state have difficulty in permeating dead membranes such as vegetable parchment paper or a dead animal bladder. This property, however, has absolutely no physiological significance and considering it without further discussion could easily lead to totally erroneous theories. In relation to the physiological effect, it is in fact the great ability of aqueous solutions of these compounds to penetrate the living tissue cells that is important. Non-specifically acting compounds undergo very rapid penetration into all living plant and animal cells. Indeed, their immediate loss from these cells following a reduction in the concentration of the corresponding compound in the surrounding medium is a condition *sine qua non*, which permits these compounds to produce their narcotic effect, despite the fact that not every compound that satisfies this condition can be considered a true anaesthetic or narcotic.

The observation of Dubois that all anaesthetics have a specific heat less than water is really superfluous, since water has the highest specific heat, and the specific heat of anaesthetics in general is lower than that of organic compounds that do not demonstrate any narcotic effect.

The assertion that the anaesthetic strength of compounds in general increases with increasing vapour pressure is entirely irrelevant. In fact, as will be shown below, exactly the opposite is true for compounds most readily compared, such as successive members of a homologous series.

1.3 INHALATION ANAESTHETICS AND OTHER NON-SPECIFIC NARCOTICS

From the purely applied perspective of the practising surgeon, there may be some justification for combining within a special group the inhalation anaesthetics that are introduced into the organism quickly and easily by means of inspired air, and can be removed by means of the expired air. A second such group could comprise those narcotics that can be introduced into the organism only in the form of a liquid or in solution. In Germany, in fact, the word anaesthetic usually seems to be employed in this sense.

Among organisms that breathe through the lungs, and especially among mammals, only inhalation anaesthetics can be removed again from the organism without operative intervention (see Appendix A: Detoxification by means of dialysis). As a result, only these inhalation anaesthetics can be employed in surgery, unless narcosis is induced by the combined action of two or more narcotics (e.g. morphine and chloroform).

From a practical point of view, there is little objection to this division of narcotics into groups. Nevertheless, it must be emphasized that, from the point of view of general and comparative physiology, the separation of inhalation anaesthetics from the other non-specific narcotics does not have the slightest justification. The mechanism of action of the inhalation anaesthetics regarding the induction of narcosis itself corresponds entirely to that of the other non-specific narcotics. In addition, they both reach their actual site of action, i.e. the cell tissue and in particular, the brain cells, from the blood in the same way. The particular pathways through which the narcotics enter the blood system are immaterial so long as they achieve the necessary concentrations in the blood. For organisms that breathe through gills, incidentally, inhalation anaesthetics reach the blood in exactly the same fashion as the other narcotics.

When such experiments are properly designed, there is no detectable difference between the two experiments relative to the time required for the effect to take place. Moreover, when the narcotized organisms are transferred from the narcotic solution into pure water, in most cases, the narcosis disappears equally quickly regardless of whether the narcosis was produced by non-volatile, non-specific narcotics or by ether, chloroform, or other volatile ones. In the few instances in which this is not the case, it was simply because the respective compounds penetrate more slowly into the blood from the surrounding solution through the epithelia of the gills and from the blood into the cell tissue, or out of the blood through the epithelia of the gills into the surrounding water and out of the cell tissue into the blood system. I will return to this last point later.

1.4 NON-SPECIFIC AND BASIC NARCOTICS

While there can be no physiological justification in distinguishing between anaesthetics and narcotics, important distinctions do exist between non-specific narcotics and basic, or salt-like narcotics.

Among the majority of compounds belonging to the second group of narcotics, there can be no doubt that it is only the basic component of the salt that is responsible for the narcotic effect, even when such compounds are introduced into the organism in the form of salts. This is certainly the case for salts of morphine and similar alkaloids. In the absence of special activity in the protoplasm of the relevant cells, most alkaloids, in the form of salts, cannot penetrate into living plant and animal cells to any extent. However, these salts are, for the most part, decomposed by the alkali of the blood (possibly the intestinal juices), and all living plant and animal cells are permeable to the free alkaloids. In fact, with few exceptions (e.g. morphine penetrates relatively slowly into the living cell), they readily permeate cells.

Reference has already been made to the fact that Claude Bernard observed no narcosis in the epithelia when he transferred ciliary epithelia into solutions of salts of morphine and other basic narcotics. Nor is the plasma flow in plant cells stopped when they are placed in solutions of morphine hydrochloride. These negative results are partly due to the fact that under these conditions only traces of free morphine, formed through hydrolytic dissociation or by the presence of small amounts of alkali, exist in these solutions. If some sodium carbonate is added to a solution of a morphine salt, or of many other alkaloids, narcosis will be produced in some plant cells. Among non-specifically acting substances, the concentrations of a given compound that are just sufficient to narcotize different plant cells, infusoria, ciliary cells, and others, hardly deviate from an average value. However, this is not at all the case with the basic narcotics. The concentrations required of a free alkaloid to produce complete narcosis vary among different plant cells, protozoa, ciliary epithelia, and other organisms by extraordinarily large amounts, frequently by more than a ten-fold factor. The same is true, incidentally, of the effect of these basic narcotics on the ganglia cells of various organisms. In frogs, for instance, complete narcosis cannot be achieved by means of morphine. By contrast, it will be shown later that the concentration of chloroform or ether in blood plasma producing complete narcosis in the frog, mammal, or even humans, varies only slightly, if at all.

The uniformity of effect of the non-specific narcotics, as opposed to the greatly variable effect of the basic narcotics already indicates a different mechanism of action of these two classes of compounds. Meanwhile, it will be shown that, as so often occurs in nature, gradual transitions exist between the two groups, however much they differ from one another in their typical characteristics.

1.5 FACTORS TO BE CONSIDERED IN DEVELOPING A THEORY OF NARCOSIS

We will mainly be concerned in what follows with the non-specific narcotics, and will attempt to explain their mechanism of action.

In order to reveal the underlying principles governing a phenomenon, it is first necessary to identify what specific factors contribute to it through their combined effect. Then an arbitrary change of one of them is made, and a determination made of the changes that are required in the other factors to compensate for it. In physiology, however, it is not always possible to identify all of the factors that contribute independently to a process. In most cases, only a certain number of such factors can be widely varied, while others are either invariant, or can be varied independently only within narrow limits.

Thus, in cerebral narcosis, the nature of the brain cells, the organism, their state of development, temperature, and nature and concentration of the narcotic play a role. Of these factors, the exact nature of the brain cells is not a simple factor, and has associated secondary considerations. There is every reason to believe that only certain brain cell constituents are closely involved in the induction of narcosis. This, of course, does not exclude the possibility that other parts of the ganglia cells are affected secondarily.

For normal organisms of a species at a common stage of development, it is reasonable to assume that the factor or factors related to the brain cells are constant. In any case, these factors cannot be varied at will. Whether or not these factors remain the same in different organisms can be determined by experiment.

1.6 RELATIONSHIP BETWEEN DOSE AND MEANS OF ADMINISTRATION

It is of the utmost importance in the study of narcosis to determine the proportions of doses of the various narcotics that are necessary and sufficient to induce narcosis in a given animal species at a certain temperature.

Before carrying out this task, a method is needed to express these relative doses. This represents a continuing question in all toxicological studies, and therefore, must be considered from a general perspective.

In the pharmacological and toxicological literature, it has been customary for some time to express the relative doses of toxins and medicines as the number of grams of compound required per kilogram of the experimental animal's body weight to produce a certain physiological response in the animal. When neutralizing a certain quantity of alkali solution in a test tube, an equivalent amount of acid must be used, regardless of whether the acid is added all at once, or in portions over a longer period of time. By contrast, it was discovered some time ago that it does make a difference whether a certain amount of a toxin or drug is administered to a human or animal in a single dose, or in several divided doses spread out over longer time intervals. Substantial quantities of even the most powerful known toxins can be taken without harm, if each ingested dose remains below a certain amount, and a sufficiently long time interval elapses between doses. In physiology, it is usually a matter of the amount of a compound per dose that is administered to an experimental animal.

Furthermore, it has been shown that the means or route of administration of a test compound can also be of great importance. For example, the same amount of a compound produces a different response, depending upon whether it is introduced into the stomach, the rectum; introduced subcutaneously into the pleural or peritoneal cavity; or introduced into the trachea or finally directly into the bloodstream of the animal. The origin of this differing behaviour is due presumedly to the fact that a certain amount of the compound produces a different response depending upon whether it is given in a single dose, or divided into several interrupted doses. Thus, a compound is absorbed into the bloodstream from different parts of the body at very different rates. For example, it is generally absorbed much more slowly from the stomach than from the pleural or peritoneal cavity. If a compound is only slowly absorbed into the bloodstream, then the molecules that are absorbed first, either by breakdown or as a result of excretion by specialized cells, will already have been removed from the bloodstream before those molecules absorbed last have entered the bloodstream. Many compounds, moreover, are absorbed only very incompletely by the stomach, and may under certain circumstances, undergo a chemical transformation before entering the bloodstream.

The discoveries mentioned, as well as many others, demonstrate that the concentration of a foreign compound in the blood, or rather the blood plasma, is of primary importance with respect to the intensity of its effect. It is desirable to know this concentration in every case, but whether this information alone is sufficient is a question that must be discussed later.

The introduction of a compound directly into the bloodstream has the disadvantage that, under these circumstances, the compound immediately comes in contact with a small amount of blood in a very concentrated form. This circumstance could easily cause blood corpuscles to be destroyed, or other changes to take place in the blood, with all their consequences. Therefore, it is customary in pharmacological experiments, except in the case of volatile substances, to inject solutions of test compounds, either subcutaneously or in the peritoneal cavity. Most of the crystalloids (Editor's note: substances capable of both crystallizing and diffusing readily across membranes when in solution) enter the bloodstream relatively quickly from these two sites, particularly from the peritoneal cavity, so that the greater part of the compound has already entered the blood before the parts absorbed have decomposed, changed, or been eliminated. With this means of administration, the maximum concentration of the compound in the blood will, after a while, be almost as high as if the same quantity of the substance has been introduced directly into the bloodstream, and equally distributed throughout the entire blood volume.

1.7 CALCULATION OF THE CONCENTRATION OF A TOXICANT IN THE BLOOD PLASMA

Is it possible to calculate the concentration of a compound in the blood plasma, knowing its dose, the weight of the test animal, and assuming the compound is absorbed very quickly?

Many pharmacologists have posed this question, perhaps in a different form. We certainly find no discussion of this in most handbooks and textbooks of pharmacology and toxicology. In fact, no general method exists for calculating the concentration of a compound from a known weight of the animal and the amount of the dose given, without further knowledge of the physical and physiological behaviour of the relevant compound.

Many years ago, Claude Bernard [3] expressed the opinion that it would be more correct to relate the weight of the toxicant administered only to the weight of the blood, rather than to the entire body weight of the test animal. He probably based this on the assumption that the relevant toxicant would penetrate only those tissue elements upon which it selectively exercises its effect. This assumption may apply more or less in some cases, e.g. for curare, and there is no doubt that, if this assumption were generally applicable and if the weight of the affected tissues constituted only a small part of the entire body weight, then Claude Bernard would be completely correct in relating the amount of the toxicant to the weight of the blood in the animal, and the task of calculating fairly exactly the concentration of the toxicant in the blood plasma would be rather simple.

In fact, however, a large number of toxicants, including all non-specific narcotics, penetrate not only the tissue in which their effect is very noticeable, but also the remaining cell tissues of the organism. In fact, this diffusion from the blood capillaries into the tissues occurs very quickly. In mammals, for example, the blood constitutes only 30% of the entire body weight, with the water in the blood comprising only 9–10% of the body water content. As a result, the concentration of these compounds in the blood plasma is greatly reduced due to this diffusion process.

1.8 CONDITIONS AFFECTING BLOOD PLASMA TOXICANT CONCENTRATION

Other conditions can also affect the concentration level of the toxicant, particularly its relative solubility in water and fats and similar dissolving agents. If a compound is able to penetrate all tissue cells and if some cells, e.g. in adipose tissue, contain a large amount of fat, then the compound will accumulate in these fats to an extent corresponding to the partition coefficient of the compound between water and the relevant fat. If the compound is much more soluble in fat than in water, and if the test animal has significant amounts of fat, then a large portion of the compound may be removed in this fashion from the blood plasma and its concentration in the blood plasma greatly reduced. This is an important consideration, since in fact, all compounds that are easily dissolved in fats are clearly able to penetrate all tissue cells [4]. Another factor to be considered is that the composition of fats varies among organisms. In humans and some other organisms, for example, triolein predominates, and at body temperature the fat is present in liquid form. However, in the case of other organisms in which tripalmitin and tristearin predominate, it is very unlikely that the amount of triolein is sufficient to keep these esters completely dissolved at body temperature. (For example, mutton fat does not become completely liquid until it reaches 60°C; it is of course possible that in the living animal, the fat is in a super-heated condition.) Thus, the fat would have a thicker consistency even while the animal is alive. These differences in the composition of fat for different organisms likewise affect the partition coefficients of the compounds between water and fats. Like fats, lecithin, cholesterol, and similar compounds that are generally distributed throughout the tissues can also remove significant amounts of the compound being tested from the blood plasma. In addition, it is very likely that absorption of the compounds occurs from both intracellular and extracellular tissue structures, which in turn would further lower the concentration of foreign compounds in the blood plasma.

A consideration of all these complications indicates that it is no simple matter to calculate even approximately the concentration of the compound in the blood plasma from the dosage given and the weight of the test animal. In order to accomplish this, a knowledge is required of (1) whether or not the compound penetrates the various tissue cells; (2) the approximate fat content of the animal and its level of lecithin, cholesterol, and other lipoids; and (3) the partition coefficients of the compound between water and the relevant fats, between water and lecithin, and between water and the other lipoid constituents.

Added to these difficulties is the fact that the concentration of the compound in the blood plasma, even if the compound is not volatile, changes partly as a result of chemical transformations, and partly as a result of excretion by specialized cells. Moreover, these changes in concentration occur at varying rates depending upon the chemical nature of the compound. A test compound which is both extremely volatile and only slightly soluble in water will be removed very quickly from the blood by the lungs with expired air. For this reason, it is hardly possible to achieve narcosis in warm-blooded animals by subcutaneous injection of aqueous or oily solutions of chloroform or ethyl ether.

1.9 BERT'S METHOD FOR MAINTAINING A CONSTANT CONCENTRATION OF AN ANAESTHETIC IN THE BLOOD

It was, however, with very volatile compounds such as chloroform and ether, which usually seem to be beset with the greatest problems, that success was first achieved in finding a method to maintain the concentration of the compound at a constant level in the blood plasma. This method (described below) avoids the complications involved in the passage of the compound into the various tissues and also those arising as a result of the differing fat, lecithin and cholesterol levels of the body. In addition, this method makes the concentration of the compound in the blood plasma independent of all other existing complications.

While this method gives no direct information on the absolute concentration of the compound in the blood plasma, it does provide a value, assuming the same conditions, that is directly proportional to the concentration.

The method consists of having the test animal or person breathe air which is mixed with vapours of the test compound under known and constant (partial or atmospheric) pressure. Under these conditions, the blood absorbs the vapours of the foreign compound until the concentration of the compound in the blood plasma reaches a defined proportion of the partial vapour pressure of the compound in the inhaled air. In order for equilibrium to occur, the concentration of the compound in the blood plasma must be directly proportional to its partial vapour pressure, so long as the temperature remains constant.

For example, suppose that on one occasion the inspired air contains 10 grams of vapour in 100 litres of air, and at another time, 20 grams. After equilibrium has been reached between the content of the compound in the blood and the respective air mixtures, the concentration of the compound in the blood plasma will be twice as great in the second case as in the first. This demonstrates that all concentration changes in the blood plasma that occur during circulation and transfer to tissue and other organism compartments are equalized again in the pulmonary circulation so long as the inspired air mixture remains the same in composition. This method is based on the application of Henry's Law, or the Dalton–Henry Law, relating to the absorption of gases and vapours by liquids. Of course, the assumption is made that the epithelia of the lungs are readily permeable to such vapours and remain strictly passive during absorption into the blood. I have confirmed for myself that this condition is actually fulfilled for all narcotizing vapours. Thus, a unit volume of the circulating blood absorbs the same quantity of compound, at a given partial pressure of the vapours, as the same volume of blood would absorb under the same conditions of temperature and partial pressure, even outside the body.

It was hardly an accident that Paul Bert first applied this method, even if its full significance was not clear to him, and could not become clear to him without some knowledge of the osmotic properties of the cell. As a result of Bert's classic studies of the significance of partial pressure with respect to the physiological action of various gases, he was stimulated to foresee a corresponding proportionality with respect to the vapours of highly volatile narcotics. Since he had already investigated the physiological effects of nitrous oxide and carbon dioxide at various partial pressures, his thoughts were all the more likely to be directed to the volatile narcotics.

1.10 BERT'S EXPERIMENTS WITH CHLOROFORM AND ETHYL ETHER

Bert's numerous studies on the effects of highly volatile compounds at constant partial pressures were published between 1880 and 1885, in *Comptes rendus des séances et Mémoires de la Societé de Biologie*, as well as *Comptes rendus de l'Academie des Sciences* [5]. Almost all of these studies are based upon experiments with known concentrations of chloroform and ether in air; some were performed using nitrous oxide.

In the following, I will provide the most important results of Bert's studies, since we will refer to them later. In his studies of chloroform narcosis, Bert found that if a dog employed as a test animal inhales air containing only 4 grams of chloroform in 100 litres, neither true narcosis, nor even just insenstivity to pain (analgesia) occurs, regardless of the duration of inhalation. Despite this, death of the animal occurs after nine or ten hours, preceded by a lowered body temperature. However, Bert did not take into consideration the fact that when the body temperature is lowered, given the same partial pressure of the chloroform vapours in the air mixture, an increase takes place in the concentration of chloroform in the blood (this is discussed on p. 98). At a concentration of 6 grams chloroform in 100 litres of air, sensitivity to pain is reduced, but not completely suppressed. Death occurs after six to seven hours. A mixture of 8 grams per 100 litres slowly produces anaesthesia and deaths in about four hours. A mixture of 10 grams of chloroform in 100 litres of air produces complete narcosis in a few minutes and death in two to three hours. For an air mixture containing 15 grams of chloroform in 100 litres of air, narcosis occurs almost immediately, and if inhalation of the air mixture continues, death occurs within about 40 minutes. Finally, an animal inhaling a mixture containing 30 grams of chloroform in 100 litres of air becomes fully narcotized after a few breaths, and dies after about three minutes.

On the other hand, if a dog inhales an air mixture containing 10 or 12 grams of chloroform vapour in 100 litres of air until complete narcosis has taken place, followed by a mixture containing only 4 grams of chloroform in 100 litres of air, the dog awakens after a while. Even if the inhaled air mixture subsequently contains 6 grams of chloroform in 100 litres of air, the narcosis will gradually be reduced.

It can be deduced from these experiments that actual narcosis is not produced in a dog until the concentration of chloroform in the blood plasma reaches the level at which the blood (at body temperature) is unable to absorb any more chloroform from an air mixture containing 8 grams of chloroform in 100 litres of air. Furthermore, narcosis continues so long as this concentration is maintained. However, when the chloroform concentration in the blood plasma falls considerably below this level, narcosis becomes incomplete or entirely disappears.

These findings demonstrate that a state of approximate equilibrium is achieved between the physiological state of the test organism and its blood plasma chloroform concentration. Several different levels of narcosis can be distinguished relative to the concentration of the narcotic established in the blood. Aside from the initial onset, the following stages of narcosis occur with increase in the concentration of chloroform in the blood plasma:

1. Decrease in, and loss of, sensitivity to pain, while partly retaining intelligence, tactile sensation and reflexes.
2. Loss of sensation of taste and loss of reflexes, and finally the loss of the reflex of the conjunctiva.
3. Complete relaxation of the musculature, followed, if the concentration of the chloroform in the blood increases further, by cessation of breathing movements and heart beat.

The blood plasma chloroform concentrations for these different stages of narcosis correspond at equilibrium to a chloroform level in the inspired air mixture of between 6 and 15 grams in 100 litres of the mixture.

Bert found the following conditions to hold for narcosis with ethyl ether: for an air mixture containing 20 grams of ether in 100 litres, narcosis takes place within about 30 minutes, preceded by severe agitation. Continued inhalation of this air mixture leads to death of the dog after about two hours and 20 minutes. If the ether level in the air mixture falls significantly below 20 grams in 100 litres, then the animal reawakens. If the ether level of the air mixture is 25 grams in 100 litres, a dog will be narcotized after 10 minutes and dead within two hours. At an ether level of 30 grams, it will die within 1 hour and 45 minutes. Finally, at an ether level of 50 grams in 100 litres, it will die within just 40 minutes.

Using this method of inhaling air mixtures of known and constant concentrations of the anaesthetic of interest does not provide, as mentioned above, any direct information regarding the concentration of the anaesthetic in the blood plasma, but indicates only that the concentration must remain constant if the degree of narcosis is to be maintained. The fact that the test animal dies after a certain testing period, even if the concentration is only just sufficient to maintain narcosis, and even if the concentration remains below this level, demonstrates that complete equilibrium is not achieved with respect to the concentration of the anaesthetic and the physiological condition of the animal. Therefore, this phenomenon must be attributed to secondary effects of the anaesthetics. In addition, Bert did not consider that in warm-blooded animals, lowering the temperature also plays a role. It causes the concentration of the anaesthetic in the blood to increase if the experiment is continued, since at a lower temperature, blood absorbs more chloroform, ether, and other anaesthetics at a given partial pressure than at a higher temperature.

Bert's method also made possible a comparison of the relative sensitivity of humans and animals with the same blood temperature to the same anaesthetic. This comparison led to the unexpected finding that humans, despite the much higher development of their brains, are not significantly more sensitive than dogs to ether and chloroform. The air mixture must contain about 8 grams of chloroform or 20 grams of ether in 100 litres in order to produce and maintain narcosis in man. Since there is little difference between the blood temperatures of humans and other mammals, the concentrations of chloroform and ether in both types of plasma must be essentially the same (I deliberately refer to the concentration in the blood plasma, and not the concentration in the blood because, in general, the concentration of a non-specific narcotic is different in the blood plasma and the blood corpuscles, due to the cholesterol–lecithin content of the blood corpuscles and other factors).

I will show later, when discussing my own experiments, how the absolute values of the chloroform and ether concentrations in the blood plasma can be calculated if the blood temperature of the test animal and the partial pressure of the vapour of the anaesthetic are known. We will then see that even frogs (i.e. tadpoles) are narcotized by about the same concentrations of chloroform and ether as man.

Experiments on narcosis with air mixtures containing a known and constant quantity of the narcotic in a fixed volume of air mixture require a rather sophisticated apparatus (at least for the experiments carried out with fairly large animals). As a result, up to now these types of experiments have been performed with only a few narcotic compounds and this method of anaesthesia has found very little acceptance in the practice of surgery, even though its advantages are indisputable.

Bert himself used a special kind of gasometer in which the anaesthetizing air mixtures are prepared and kept and from which the air mixture is administered to the test subject, human or animal, by means of a tube and mask (in the case of animals, sometimes also by means of a tracheal tube). Special valves ensure that with each breath, a fresh portion of the air mixture is inhaled from the gasometer, while the air previously inspired exits upon being exhaled. Since the air mixture cannot be displaced from the gasometer by means of water, which absorbs large amounts of chloroform and ether, the space in the gasometer needs to be decreased by displacement of the upper half into the lower half. A

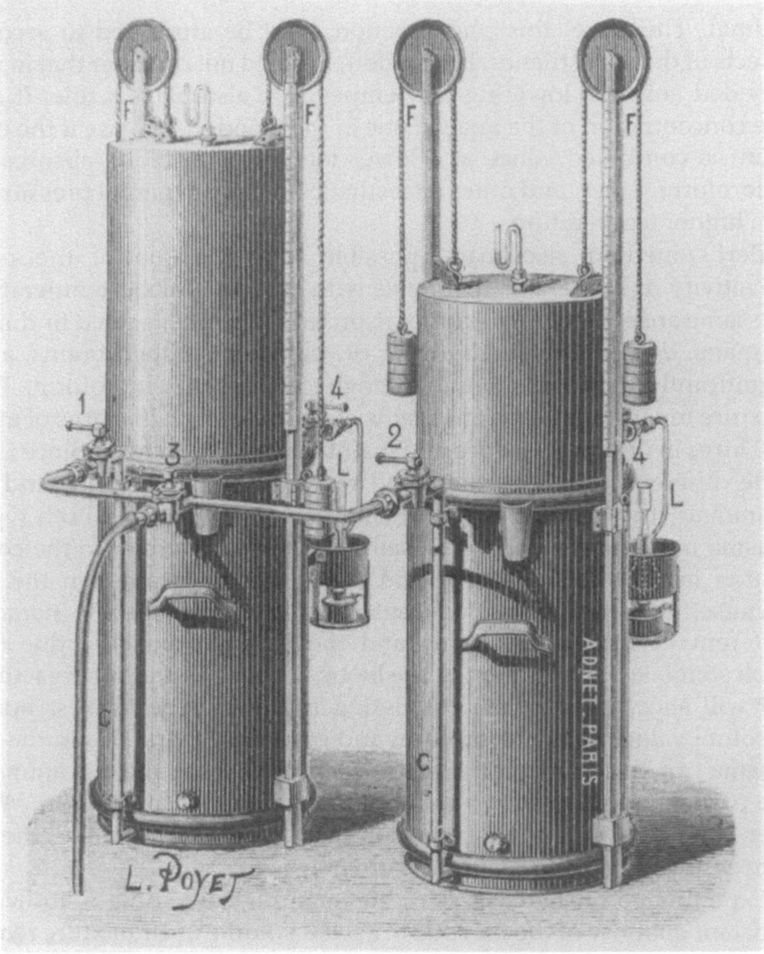

Fig. 1.1 Double gasometer according to Saint-Martin for stoichiometric mixtures. (Reproduced from Dastre, 1890.)

more detailed description of this gasometer can be found in Dastre (1890, *Les Anesthésiques*, p. 105) (Fig. 1.1).

An ingenious machine that can produce air mixtures with a specific chloroform content at the same rate as they are consumed was constructed by R. Dubois (Fig. 1.2) (*Compt. rend. des séances et Mém. de la Soc. Biol.*, June 14, 1884 and 1885; in Dastre, *Les Anesthésiques*, p. 108, and in R. Dubois, *Anesthésie Physiologie*, 1894, p. 107 and 109). Although this machine is relatively effective and has produced very favourable results,

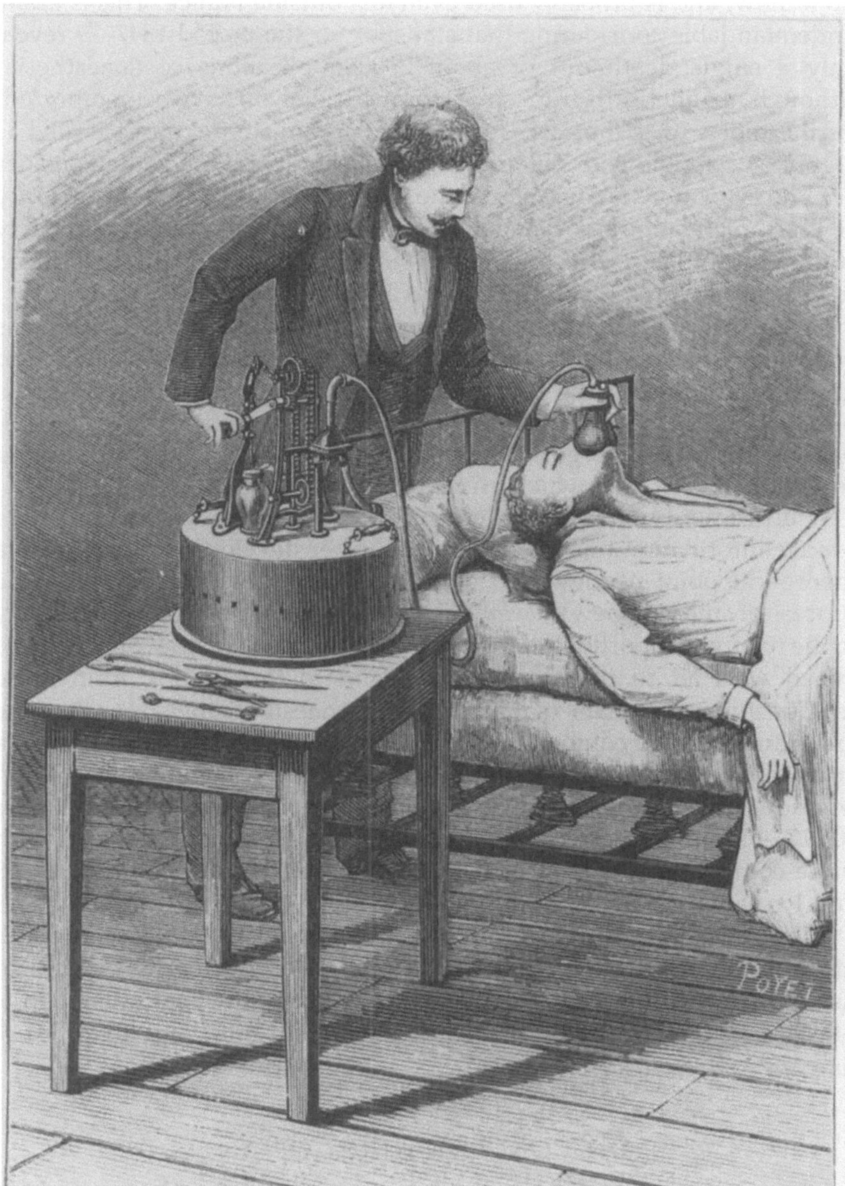

Fig. 1.2 Apparatus of Dastre for producing stoichiometric mixtures of anaesthetic and air. (Reproduced from Dastre, 1890).

even in France it has met with very little acceptance in surgical practice up to now, and is probably used even less outside France. This is easily understandable considering that statistics for the period 1847–79 reveal only a single death out of about 3000 applications of anaesthesia, although not all deaths from chloroform are likely to have been reported. Furthermore, once the earlier method of administering air saturated as highly as possible with chloroform had been abandoned, the statistics became even more favourable for the most recent period. In the majority of cases of fatality associated with chloroform, the patients probably suffered from heart diseases and it is unlikely that anaesthesia induced by precise mixtures of chloroform and air would have prevented death.

Given this state of affairs, it is unlikely that anaesthesia by means of air mixtures of known and constant anaesthetic levels will find much application in the future outside physiology and physiological psychology. In these fields, however, this method may be very useful, particularly in the closer study of the psychologically very interesting condition of excitation which precedes narcosis.

While these studies of Paul Bert signify great progress in gaining more precise information on narcosis, their further development has, from a theoretical point of view, led along quite incorrect pathways. In the process of studying new anaesthetics, attention has been paid exclusively to the relative proportions of the corresponding anaesthetics needed to be mixed with air to produce narcosis, and to formulating from these relationships as a means of measuring the anaesthetizing strength of the individual narcotic compounds. Thus, Dastre tells us [6] that the relative anaesthetic strengths of the four compounds ethyl ether, amylene, ethyl chloride, and chloroform can be expressed, respectively, by the numbers 20, 17, 13 and 10, because, in order to achieve complete narcosis with these compounds, 100 litres of air must contain respectively, 20, 17, 13 and 10 grams of these compounds.

It is obvious that this means of measuring the relative narcotic strength of a compound is arbitrary and inappropriate. It is much more practical to use as a measure of the relative narcotic strength the lowest partial pressures of the various anaesthetics that are sufficient to produce narcosis.

1.11 CONCENTRATION OF AN ANAESTHETIC IN THE BLOOD PLASMA

Of primary importance is the concentration of the anaesthetic in the blood plasma, the intercellular lymph, and the fluid surrounding the ganglia cells. The concentrations at these three sites should be equivalent for anaesthetics, as will be shown later. The goal is not to determine the pressures of the various anaesthetics in the air required for narcosis, but to find an agent that will produce the desired concentration in the blood plasma, and maintain it at a constant level. However, the partial pressure in the air mixture and the concentration in the blood plasma are only in a constant ratio for the same anaesthetic. By contrast, the relative proportions with respect to these two amounts are completely variable for different anaesthetics and depend upon the solubility of the individual anaesthetics in water (or in the blood plasma). Even for the same anaesthetic, this ratio remains constant only for a certain blood temperature and it changes considerably with temperature, a factor which must be taken into consideration when performing comparative narcosis experiments on cold- and warm-blooded animals.

Thus, if a frog and a mammal are placed together in the same vessel containing ether vapours and if the temperature is maintained constantly at about 15°C, equilibrium can be established between the ether concentration in air and its corresponding blood plasma concentration in each organism. Under these conditions, the concentration of the ether in the blood plasma of the frog will be about three times greater than in the blood plasma of the mammal (whose temperature is about 38°C). This reflects the fact that at a given partial pressure, analogously to the absorption of gases, water and other liquids absorb more of the vapours of the compound at a lower temperature of the relevant liquids. At this point it should be emphasized that, even with gases, it is their molar concentrations in the blood plasma or their corresponding partial osmotic pressures in the blood plasma that are the physiologically significant values and not the partial pressures of the relevant gases in the inspired air. This fact should never be overlooked when doing comparative physiological experiments. Following Bert, it is customary, when conducting physiological experiments on the effects of gases, to express ratios quantitatively by means of the corresponding partial pressures. Thus, the statement that oxygen is more toxic than carbon dioxide, although entirely true, appears to be a paradox. Later, I will discuss how the concentrations of ether and chloroform in the blood plasma of the test animal can be calculated from their partial pressure in the inspired air.

In the literature on chloroform or ether narcosis there exists an erroneous concept which needs to be discussed briefly here. Frequently, the sensitivity of different organisms to these two anaesthetics is determined according to the length of time required for narcosis to occur at a given concentration in air of the inhaled anaesthetic. Dastre [7] reported that:

> In one and the same container, organisms of varying organizational levels, e.g. a bird, a mouse, a frog and a sensitive plant (*Mimosa*

> *sensitiva*) are exposed to ether vapours. The bird, which possesses a more delicate organization and a greater vitality, was found to stumble and fall down insensibly after four minutes. The mouse showed no further signs of sensibility after ten minutes. The frog became paralysed later, and the plant did not become insensitive to stimuli until after 25 minutes.

From this chronological succession of events, it was concluded that the bird is more sensitive to ether than the mouse, the mouse more sensitive than the frog, and the frog more sensitive than the plant. This form of inference, which has been employed by many other authors in one form or another, is totally unjustified and misleading. Aside from the fact that the temperature must always be taken into consideration when comparing cold-blooded animals and plants with warm-blooded animals, another factor must be considered. The rate at which the blood plasma becomes saturated with the ether vapours in the air at a certain pressure, or rather reaches a certain fraction of the saturation value, is dependent upon the amount of blood and the fat level of the test animal and on many other factors. One species of animal can be accurately characterized as being more sensitive to a certain compound only if a lower concentration of the compound in the blood plasma of the first animal is sufficient to produce the same physiological effect as in the second animal.

1.12 THE INTERCELLULAR LYMPH AS A PATHWAY BETWEEN THE BLOOD AND THE TISSUE CELLS

Up to this point, our attention has been focused almost exclusively upon the relative concentrations of foreign compounds within the blood plasma. We have examined only incidentally the question of their concentration within the intercellular lymph and tissue cells, as well as the effect of test duration on the response to foreign compounds.

The cell tissues, which are nearly always the actual sites of action of foreign compounds within the organism, are not directly surrounded by the blood plasma, but by the intercellular lymph. The intercellular lymph has the same relationship to cell tissues as that of a nutritive solution to water algae or fungi growing within it. The cell tissues not only accumulate their normal nutrients from this intercellular lymph, but also absorb any foreign compounds, just as they also release into it their metabolic transformation products. Meanwhile, there is some foundation for the assumption that, within a short period of time, the concentration of all crystalloid* substances introduced into the blood reach essentially the same level in the intercellular lymph as in the blood plasma. This is

* Editor's note: Crystalline substances that are soluble in water.

always the case with those compounds that can both rapidly penetrate all tissue cells, and be eliminated, since they can be transported directly through the protoplasm organelle of the capillary endothelia. In the case of crystalloid substances whose solutions are unable to penetrate, or penetrate only slowly, the living cells by pure osmosis, the equalization of the concentration in the blood plasma and the concentration in the intercellular lymph may take place much more slowly. This probably reflects the fact that the only path available to these compounds is through the cement-like substance of the endothelia and its stomata (assuming that these are actually present in the normal condition of the capillaries), unless these compounds are absorbed from the blood plasma as a result of the activity of the capillary epithelia and secreted again into the lymph of the spaces in the tissues. Unfortunately, hardly any quantitative studies have been performed of the transport of foreign compounds from the blood plasma into the lymph. This uncertainty in our knowledge of this distribution is limited, however, only to those substances for which penetration of the living cell takes place either very slowly, or not at all. Fortunately, for our research, the non-specific narcotics and many other toxicants do not fall within this category.

1.13 THREE GROUPS OF COMPOUNDS DIFFERING WITH RESPECT TO THEIR PERMEABILITY TO TISSUE CELLS

A moment ago, I drew a parallel between the relationship of cell tissues to the surrounding lymph, and the relationship of water algae and fungi to the surrounding nutritive solution. This analogy remains valid for a compound introduced into the blood and intercellular lymph, or within the nutritive solution of the algae. Any soluble crystalloid compound added to the algal nutritive solution will permeate the cellulose membrane, and reach the outer plasma membrane. Beyond this point, however, different compounds behave differently. Substances can be subdivided into three groups, according to their ability to distribute themselves within the protoplasm in proportion to their rate of distribution. Moreoever, exactly the same three cases can be distinguished for foreign compounds present within the intercellular lymph, with respect to their behaviour towards the cell tissues.

1.13.1 Compounds unable to penetrate living cells

We can place within the first group all compounds which are unable to penetrate the interior of the protoplasm, so long as the algae are alive and remain unharmed. These compounds therefore act primarily only on the

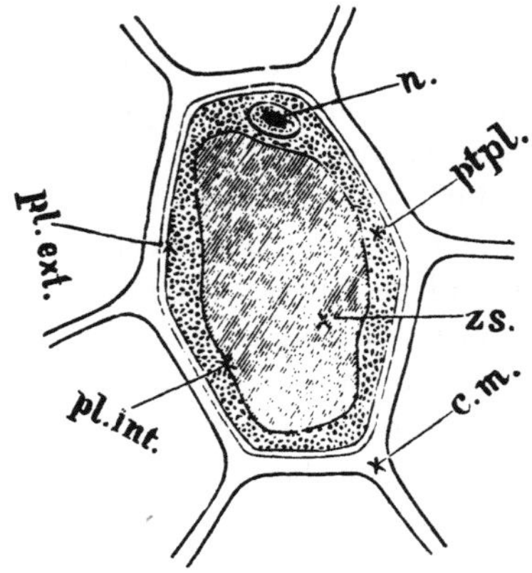

Fig. 1.3 Representation of a plant cell. c.m., cellulose membrane; ptpl., protoplasm; pl. ext., external boundary layer (external plasma membrane); pl. int., inner boundary layer (vacuole membrane); n., nucleus; zs., plasma fluid. (Reproduced from Overton, 1895.)

outer interfacial layer of the protoplasm (Editor's note: see Fig. 1.3 for Overton's representation of a plant cell showing these structures.) . After a sufficient period of time, the protoplast can, of course, be damaged or killed by this purely external effect, resulting in the penetration of the compound into the interior of the protoplast. The action of potassium dichromate and many other acid salts represent good examples of this effect. In these cases, the concentration of the foreign compound naturally also plays a very important role and is the factor that determines whether such a compound has a detectably harmful effect, and whether the effect occurs within a short or long period of time.

1.13.2 Compounds that readily penetrate living cells

Within the second group, we can place compounds that not only immediately penetrate the cellulose membrane, but also encounter no barrier in the outer plasma membrane, i.e. the outer interfacial layer of the protoplast. They penetrate the interior of the protoplast without hindrance, completely impregnate the protoplasm, reach the inner surface layer of the protoplasm (inner plasma membrane, vacuole membrane), permeate this also, and thus finally reach the cell fluid. The concentration of these compounds within the cell fluid increases very quickly until they reach about the same level as in the outer fluid. (Note that if the solutions of a compound in pure water and in a salt solution are separated by a semi-permeable wall which is permeable to the relevant compound, but not to the salt molecules, a state of equilibrium is reached, not at equal concentrations of the compound on each side of the wall, but at a lower concentration of the compound in the salt solution. The differences in concentration can be very significant for a concentrated solution, which is usually not the case with cell fluids.)

This second group possesses particular significance for us because almost all non-specific narcotics, including most basic narcotics, the non-specific antipyretics, and many antiseptics belong to it. Of the non-specific narcotics, only chloralose does not belong to this group. Therefore, I will discuss this second group in somewhat greater detail. For compounds within this group, the concentration within the cell plasma almost immediately reaches essentially the same level as in the outer fluid. Nevertheless, this condition does not necessarily apply to the protoplasm itself, if the concentration of a compound in the protoplasm is defined as the weight of the compound in the combined volumes of the protoplasm and its adjuncts. For example, if the protoplasm contains a fatty or volatile oil, or any other substance with a similar dissolving capacity as one of these oils, the foreign compound will accumulate in these cellular constituents. It will accumulate until concentrations are reached that are in the same proportion to the concentration of the foreign compound in the outer fluid and in the cell fluids as the partition coefficients of the compound between the specific oil-like cell components and water. Lecithin and cholesterol appear to be present in all living plant and animal cells, and under certain circumstances, these compounds possess a very similar capacity to dissolve fatty and volatile oils. For these and other reasons, a short discussion is needed here.

Physical properties of lecithin and cholesterol

At room temperature, pure crystalline cholesterol is a poor solvent, and can dissolve few compounds. On the other hand, lecithin, which swells up in water or aqueous solutions, can, in this swelled state and even at room temperature, dissolve all compounds that are soluble in oils, with the formation of a kind of stable solution. Furthermore, a lecithin and cholesterol mixture formed by evaporation of a benzene solution, and then mixed with water, swells up in a similar way to pure lecithin, and the cholesterol remains (if not present in too high a proportion) dissolved in the swollen lecithin. This swollen mixture of lecithin and cholesterol now has a similar dissolving capacity, even when cold, to swollen lecithin alone, and absorbs significant quantities of the dissolved substance from aqueous solutions of those compounds that are more readily dissolved in ether and oil than in water. Even certain compounds that are insoluble in oil and ether, e.g. most salts of aniline dyes, are dissolved in abundance by this swollen mixture of cholesterol and lecithin (and by pure lecithin), or removed from the aqueous solutions of these dyes [8].

It is very likely that the lecithin and cholesterol of living plant and animal cells exists in the form of such a swollen mixture as this. Since this mixture has only a limited swelling capacity, and can hardly absorb all the aqueous content of the protoplasm, there is a question of how it is distributed in the protoplasm. This question is in general closely related to that of the detailed structure of the protoplasm. Although individual theories on this subject differ rather widely, the living protoplasm, can, with little objection, be compared to a completely soaked sponge. This comparison does not apply to the osmotic behaviour of cells, which is not under discussion at the moment. The honeycomb of walls and the fibres of the sponge may be considered analogous to the framework of the protoplasm. The solution held within the interstices of the honeycomb as well as between the individual fibres of the soaked sponge corresponds to the plasma fluid of the protoplasm. The latter contains compounds including potassium phosphate, fructose, and unstructured protein. It should not be assumed, of course, that the framework of the protoplasm possesses the rigidity and immobility of the honeycomb walls of the sponge. In some respects, a better comparison would be foam. All comparisons of this kind will be found imperfect if they are examined in too much detail. If we accept this concept of the structure of the protoplasm as valid, it is likely that the mixture of lecithin and cholesterol provides the building blocks for the actual protoplasm framework, as in the case of the organized protein constituents. This would particularly be the case within the protoplasm layers bordering the outside of or adjacent to a vacuole. It is very likely that the general osmotic properties of the cell depend upon the presence of this swollen lecithin–cholesterol mixture, since all compounds that are soluble in this mixture indeed penetrate living cells. Also, the rate of permeation of the compound is closely related to its relative solubility both in water and in this mixture. If this lecithin–cholesterol mixture is truly part of the protoplasmic structure, i.e. serving as a component of the protoplasm framework, then, after equilibrium has been achieved, the concentration of the compounds that readily penetrate the living cell would be nearly the same within the protoplasmic fluid as in the circumambient solution, while their total concentration in the protoplasm (i.e. their weight per unit volume of protoplasm) could be greater than in the cell-free solution which corresponds to the concentration of the compound in the plasma.

1.13.3 Compounds that slowly penetrate living cells

Finally, within the third and last group can be placed all compounds occupying an intermediate position between those in the first and second groups with respect to their penetration of the living cell. Although these compounds penetrate the living protoplasm, they do so less quickly than those of the second group. As a result, their concentration in the intracellular water of the protoplasm and the cell fluid does not reach the same levels as in the outer fluid until after some time. The time required for this process can vary from five minutes to several days for different members of this group. The full concentration of these compounds therefore acts first only on the outer surface layer of the protoplast, and even here, only from one side (Fig. 1.4). The direct effect of such substances on the internal constituents of the protoplast only gradually reaches a higher level. Among the non-specific narcotics, chloralose belongs within this group; in constrast, morphine falls within the basic narcotics. The diluted solutions of the strong mineral acids and of many polyvalent organic acids belong here as well. The effects of these acids, however, are complicated by the fact that only molecules of the undissociated acids can penetrate into the unharmed protoplast. The behaviour of strong acids is of pharmacological interest only with respect to understanding their local effects.

Having explained the relationship of the concentration of a foreign compound in the blood plasma to its concentration in the intercellular lymph, and in the intracellular water of the cell tissue, we will return to the problem of maintaining a constant concentration of compound in the blood plasma. As I have already mentioned, Bert's method represents the first successful attempt to solve the problem, although neither Bert himself, nor his followers determined the absolute concentration in the blood plasma of the compounds they tested. A particular advantage of this method is its applicability to humans and animals, which is the most

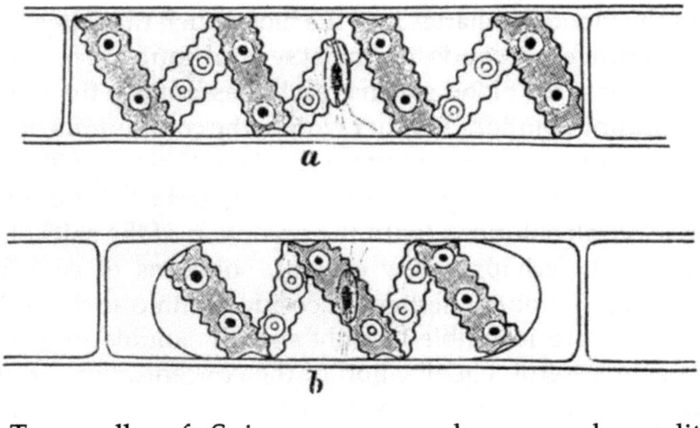

Fig. 1.4 Two cells of Spirogyra: a, under normal conditions; b, plasmolysed. (Reproduced from Overton, 1895.)

important objective of pharmacological experiments. The disadvantages of the method are that it can be used only with compounds that have a significant vapour pressure at normal temperatures, and are not too soluble in water. If these conditions are unfulfilled, too much time is required before equilibrium is reached between the vapour of the compound and the level of the compound in the blood plasma for this method to be advantageously employed.

1.14 METHOD OF PRODUCING KNOWN AND CONSTANT CONCENTRATIONS OF NON-VOLATILE COMPOUNDS IN THE BLOOD: LIMITS OF APPLICABILITY

Now the question is whether or not, at the expense of losing some of the advantages of Bert's method, another experimental method can be found that can also be used with compounds that are not very volatile at normal temperatures, or that are more soluble in water. There is a suitable experimental method that, while not applicable for all compounds, can be applied with the non-specific narcotics, and has at least a much more general application than Bert's method. Its application is, of course, limited to test animals whose respiration occurs through the skin or through gills, at least if one wants to use the method in its simplest form. It cannot be denied that this is a considerable disadvantage.

This method consists of simply placing experimental animals (tadpoles, fish, crustaceans, bryozoae, worms, and other aquatic organisms) in an aqueous solution of the compound of known concentration and choosing a high enough volume of the solution so that the concentration of the compound cannot undergo a significant change as a result of its absorption by the organisms. This method is useful only for compounds that pass so quickly through the epithelia of the gills, the skin, and the epithelia of the blood capillaries into the blood such that after a relatively short time (from a few seconds to at most several hours) the concentration of the compound in the blood plasma reaches essentially the same level as in the surrounding solution. In other words, the compounds must belong to either our group two or group three. This method is applicable to only a certain number of the compounds belonging to the third group.

In the case of salt solutions (with the exception of the salts of the basic aniline dyes and perhaps a few others), solutions of carbohydrates, alcohols of a value of four or higher, glucosides, amino acids, and all other compounds that are insoluble or only slightly soluble in a mixture of lecithin and cholesterol, equalization of the concentrations in the blood plasma and the outer solution never occurs as long as the animal is still alive. Accordingly, this method cannot be used with these compounds or at least does not provide the desired results. If, for example, tadpoles are placed in 0.5% solutions of various potassium salts, the toxic effect of these salts will usually not be apparent for a day or several days, and there is no likelihood of complete equalization between the concentration of the potassium salt in the outer fluid and in the blood plasma. The potassium salt will probably be absorbed only by the intestinal tract, and will not be absorbed, or will be hardly absorbed, by the epithelia of the gills, or by the skin.

Therefore, a condition for the application of this method is a knowledge of the osmotic properties of the epithelia of the gills and the skin. However, the experiments carried out by means of this method, if they are sufficiently varied, can themselves often provide information on these osmotic properties, as will be shown later. This method has been used for several years by Meyer and his students in the study of pharmacological questions; the author himself has used it for ten years. Both Meyer and the author have used amphibians as test animals, with a preference for tadpoles.

Before going on to discuss the experiments carried out by this method and the theoretical findings that resulted from them regarding the mechanism of narcosis, and some physiological conditions related to narcosis, it seems appropriate to give an overview of the more important hypotheses proposed up to now on the occurrence of narcosis and to undertake a critical discussion of these hypotheses.

REFERENCES

1. Dastre, A. (1890) *Les Anesthésiques. Physiologie et Applications Chirurgicales*, Paris, G. Masson, p. 158.
2. Dubois, R. (1894) *Anesthésie Physiologique et ses Applications*, George Carrbe, Paris, p. 16.
3. Bernard, C. (1857) Leçons sur les effets des substances toxiques et medicamenteux, Lesson 22, pp. 334–7.
4. Overton, E. (1899) Ueber die allgemeinen osmotischen Eigenschaften der Zelle, ihre vermutlichen Ursachen und ihre Bedeutung für die Physiologie. *Vierteljahrsschr. Naturforsch. Ges. Zürich*, **44**, 88–135.
5. Bert, P. Sur la zone maniable des agents anesthésiques et sur un nouveau procédé de chloroformation. *C.R. des Séances et Mém. de la Soc. de Biol.*, March 13, 1880, and *C.R. des Séances de l'Acad.*, Nov. 14, 1881; Méthodes d'anesthésie prolongée par des mélanges dosés d'air et de vapeur. *C.R. des Séances et Mém. de la Soc. de Biol.*, June 16, 1883; Anesthésie par l'éther. *Ibid*, Aug. 4, 1883; Application à l'homme de la méthode d'anesthésie chloroformique par les mélanges titres. *Ibid.*, Dec. 22, 1883 and Jan. 5, 1884; Sur les mélanges titrés

d'éther et d'air. *Ibid.* Mar. 1, 1884; Étude analytique de l'anesthésie par les mélanges titrés de chloroforme et d'air. *Ibid.*, July 4, 1885; and other studies.

6. Dastre, A. (1890) *Les Anesthésiques: Physiologie et Applications Chirurgicales*, G. Masson, Paris, p. 161.
7. Dastre, *Ibid.*, p. 33.
8. Overton, E. (1900) Studien über die Aufnahme der Anailinfarben durch die lebende Zelle. *Jahrb. fuer Wiss., Botanik*, **34**, 669–701.

2

Critical review of the major hypotheses on the mechanism of narcosis

2.1 HYPOTHESES BASED UPON THE CIRCULATION IN THE BRAIN

The many similarities between artificially induced narcosis and natural sleep are so obvious that it is not surprising that one of the hypotheses first proposed and still frequently mentioned attributes the same common mechanism to both conditions. Since ancient times, it has been assumed that normal sleep is associated with certain conditions of the circulation of the central nervous system and particularly the brain. Throughout antiquity and the Middle Ages and even in the first half of this century, it was believed that the condition of sleep was accompanied by an unusual accumulation of blood in the brain, that compressed its constituents, causing the activity of the brain to be interrupted or reduced. In short, sleep was thought to be induced by a hyperaemia of the brain. This theory appeared most closely in accord with the general experience that it is easier to fall asleep lying down than standing up, since hydrostatic laws meant that the blood would be more likely to accumulate in the head if the body were in a horizontal position. (Up until Claude Bernard, almost nothing was known of vasoconstrictors and vasodilators.) It was not until 1860 that this theory of sleep was disproved by the physician Durham [1] who showed by experiments on dogs that during sleep, the brain is no richer in blood than in the waking condition. In fact, sleep is characterized by a lack of blood and a consequently lowered volume, whereas when the animal awakens, the blood vessels of the brain expand, and the brain increases in volume. Durham observed the injected condition of the brain (or rather the cortex) through a hole that he had made in the cranial cap with a trephine. Later, Hammond [2] published similar observations that he had made in a man whose brain had been exposed to a considerable extent (three inches in one direction and six inches in the other) as a result of a railway accident. These observations and experimental results have subsequently been confirmed several times, and anaemia (at least physiological anaemia of the brain during sleep) can be accepted as a definite fact (Recently several authors, e.g. Hill, have denied that the brain contains noticeably less blood during sleep than when awake; the blood is said to accumulate in the veins at the base of the brain. Since, however, for the question being discussed, we are only referring to the amount of blood that flows through the cerebral cortex during a certain period of time, and there is no doubt that this is less during sleep than when awake, one may at least refer to a physiological anaemia during sleep).

In the same year that Durham published his experimental results, the American physician Bedford-Brown [3] made the observation in a patient with a skull fracture that the brain also becomes low in blood during sleep induced by an anaesthetic after becoming temporarily hyperaemic when the anaesthetic is first administered. Several researchers subsequently set up experiments with animals in order to examine the question of the circulatory condition of the brain during narcosis. At first, however, these experiments led to contradictory results, since one researcher detected anaemia and another hyperaemia during narcosis. According to Claude Bernard [4], who repeated these experiments in a critical fashion, at the onset of narcosis, while the animal is still in a state of agitation, there is normally a hyperaemia of the brain. This is very obvious from the fact that the brain substance pushes out like a hernia through the opening made by cutting a section out of the skull and the dura mater before the start of narcosis, to facilitate closer observation of the brain surface; just as occurs in a non-narcotized animal that is in a state of excitement. Soon, however, this condition is replaced by a lack of blood in the brain, resulting in the return of brain tissue that had pushed out through the opening of the skull. This tissue even withdraws from the roof of the skull, while the colour of the brain surface becomes significantly paler than in the normal waking state. However, the anaemia of the brain during artificially induced narcosis does not appear to go as far as during normal sleep (at least, this seems to be the case according to Claude Bernard's description of the phenomena). If the operation with the trephine is not carried out sufficiently skilfully, then inflammation may occur, and this may have caused some researchers to report a hyperaemia of the brain during narcosis, and to regard this erroneously as a condition peculiar to narcosis. At least for chloroform narcosis, it can at present be considered a definite fact that the cerebral cortex is low in blood in this condition.

It was natural to hypothesize that artificial narcosis is caused by a relative lack of blood in the brain, or constriction of the cerebral arteries, in the same way that this was assumed for natural sleep. According to this hypothesis, the vasomotor centres that control the blood vessels of the brain would be the actual site at which narcotics exert their effect.

This hypothesis, which has been made over and over again by researchers untrained in pharmacology, is completely untenable. This can be easily demonstrated by experiments on amphibians, in which the brain remains capable of functioning for some time even after complete suppression of brain circulation. Nevertheless amphibians can be narcotized as easily as mammals. In fact, as we will see later, the concentration of many non-specific narcotics that is just sufficient to produce complete narcosis is almost the same for amphibians and humans, and the same is true for insects and crustaceans which do not possess any blood vessels in the brain but only a kind of intercellular lymph. Moreover, in insects the blood plays a subordinate role as a vehicle for respiratory gases, since the ends of the tracheal branches often lead directly into the cell tissue. Comparative pharmacological experiments leave absolutely no doubt that narcotics (perhaps with the exception of sodium bromide and other compounds with similar effects, if these are to be included at all with the narcotics) attack the ganglia cells of the brain directly.

2.2 HYPOTHESIS OF CLAUDE BERNARD

Claude Bernard [5] proposed a hypothesis for the action of narcotics that is worthy of consideration. After discussing the various physiological means in addition to actual narcotics by which anaesthesia can be induced, such as anaemia, asphyxia and heat, he postulated that in spite of the great variety of these methods, the actual mechanism of anaesthesia always remains the same and that all these methods produce the same modification in the ganglia cells. According to him, this modification consists of a semi-coagulation of the protoplasm of the nerve cell, a coagulation that is only transient in that the protoplasm returns to its original condition after the narcotic has been removed. He stated the following (original in French):

> In our view, this action would consist of a semi-coagulation of the substance itself of the nerve cell, a coagulation that would not be permanent, i.e. the substance of the anatomical part would return to its former normal state after elimination of the toxic agent.

His hypothesis is based solely on conclusions drawn by analogy, namely from observation of the rigidity apparent in muscle fibres when the muscle has been exposed to chloroform vapours. Among other things, Claude Bernard paid insufficient attention here to quantitative relationships. I believe he was mistaken in assuming that all narcosis is based upon a common mechanism. Quite apart from narcosis produced by heat (heat numbness), there can hardly be any doubt that the mechanism of narcosis induced by non-specific narcotics is quite different from that induced by many basic narcotics. Bernard's hypothesis of a partial coagulation, perhaps expressed better as a state of greater rigidity of the protoplasm of the ganglia cells during narcosis, might apply to the narcosis produced by most of the basic narcotics. It undoubtedly does apply to a condition produced in plant cells and many animal cells by resorcinol and other hydroxybenzenes. This effect can be characterized as a narcosis of the relevant cells in so far as all protoplasm movement and other activities of the protoplasm cease, only to start again after these compounds have been removed, as after genuine narcosis. These compounds differ from actual narcotics in that they do not paralyse the ganglia cells at lower concentrations than are required for the other cell tissues, as is the case for genuine narcotics. We will return to this in the experimental section of this book.

2.3 HYPOTHESIS OF BINZ

Binz proposed, apparently independently of Claude Bernard, a theory of narcosis that essentially agrees with Claude Bernard's hypothesis. In his *Lectures on Pharmacology* (Lecture X, pp. 175–8), Binz refers first to the following 'experimental facts that explain the occurrence of the narcotic effect'. When placed in 1% morphine hydrochloride, sections of the cerebral cortex of the rabbit become dull, the centres look dusty, and all the outlines are sharper than in 1% sodium chloride solution. A similar darkening is said to occur when fresh brain cells are exposed to chloroform vapours or to a solution of neutrally-reacting chloral hydrate. Binz then continues:

> The whole process gives the impression of a coagulation narcosis, as can also be perceived when neutrally-reacting protoplasm toxicants take effect on large transparent infusoria, i.e. first quite mildly, then more strongly. In the beginning, the protoplasm becomes a little dark and the movements become slow. Later, the protoplasm becomes granulated, and the movements cease. Recovery can occur at the first stage if the toxicant is flushed out again, but not in the last stage. I compare the former with sleep, the latter with the death of

the cell. The first onset of coagulation can fade, but coagulation itself cannot be reversed.

The experiments on which Binz based his theories on the occurrence of narcosis are, however, much too crude to be able to carry very much weight. On the one hand, the quantitative relationships in his experiments are too far removed from those that prevail in normal narcosis, and on the other hand, no account is taken of normal osmotic conditions. As an example of the latter point, a 1% solution of morphine hydrochloride is not iso-osmotic with a 1% sodium chloride solution, but has the same osmotic pressure as only a 0.15% sodium chloride solution. Sections of fresh brain will absorb a great deal of water from a morphine hydrochloride solution, and as a result of this process, the ganglia cells will quickly be damaged and killed. It would be more appropriate to dissolve the morphine hydrochloride in a 0.7% or 0.8% sodium chloride solution instead of in distilled water, but even then the experiment would still be too crude for it to be of much value. As far as the experiment with chloroform vapours is concerned, it should be noted that it would be considered more significant only if the brain sections had been subjected to chloroform vapours of a quite definite and constant pressure, which was not the case, however, with the Binz experiments. In order for the experiment to correspond in some measure to the conditions of normal chloroform narcosis, it would be necessary to expose the brain sections to the vapours of a solution of chloroform in water (or preferably a 0.85% sodium chloride solution) that contained about one part by weight of chloroform to 6000 parts water. As we shall see later, the partial pressure of the chloroform vapours in such a solution corresponds to the pressure of the chloroform vapours in the blood plasma of the animal that has been narcotized completely, but not too deeply, by chloroform, assuming the temperatures of the sodium chloride solution and the blood plasma are the same.

2.4 HYPOTHESIS OF DUBOIS

A newer hypothesis for narcosis, which can be considered a modification of Claude Bernard's theory, was proposed several years ago by R. Dubois and has received quite wide acceptance in France. According to Dubois, anaesthetics cause a partial dehydration of the protoplasm, or as Dubois puts it himself, 'they increase the dissociation pressure of the imbibing water of the tissues' (Editor's note: translated from the original French as quoted by Overton). As a result, the organisms that are exposed to the effects of these dehydrating agents are removed into a state of latent life, similar to that of a dehydrated wheat kernel or desiccated conifer.

Before moving on to a critique of this hypothesis, I want to include a description in his own words of the principal experiment upon which Dubois bases his theories on the mechanism of narcosis [6] (Editor's note: quoted by Overton in French):

> If fleshy-leaved plants such as echeverias are placed in a closed container in contact with ether vapours, after a certain length of time, large droplets of water will be exuded through the cuticle. Oranges that have been kept for a while in this type of atmosphere have the appearance of frozen fruit that is thawing.
>
> It is interesting to see that the intimate action of general anaesthetics is very similar to that of cold which also inhibits absorption, causes separation of water and protoplasm in the frozen tissues, and like ether, removes the haemoglobin from the blood, etc. Cold is, of course, a well-known anaesthetic and antiseptic.

In this instance, Dubois interpreted the phenomena quite incorrectly. Even when Dubois' experiments were known to me only through reference, I immediately speculated that the secretion of water described in anaesthetized plants is unrelated to narcosis as such, but must have been the result of more extensive injury to the plants which caused the cells, or a part of the cells, to lose their normal osmotic properties, and become permeable to the compounds dissolved in the cell fluid. It is known that the cell membranes of the living plant cells under normal conditions are in a state of elastic tension since the cell content exerts a pressure of several atmospheres (usually 4–8) against them. This pressure of the cell content is of osmotic origin, and exists only so long as the protoplasts remain impermeable to the compounds dissolved in the cell fluids. If the cells die or become severely damaged, whether from exposure to too great a temperature, or to freezing, or from the effect of toxicants, then the protoplasts lose their quality of non-permeability to the dissolved components of the cell fluid, either immediately, or within a short time. Subsequently, the pressure of the cell contents against the membranes is increasingly reduced, the elastically stretched membranes contract, and some of the cell fluid is forced out through the cell membranes. In Dubois' experiments, the exuded water came out of the cell fluids, and not, or only to a minute degree, out of the protoplasm. When plant cells are exposed to the saturated vapours of ether or chloroform; they absorb far more ether or chloroform than is required for narcosis. However, at concentrations that are only half as great as the concentration required to produce complete narcosis, most plant cells are killed after a few minutes and lose their normal osmotic properties.

Dubois' experiment can be modified so that the plants are not exposed to the vapours of pure liquid ether or chloroform, but to the vapours of aqueous solutions of these two compounds at concentrations (1.25–1.5% by weight) that are just enough to produce narcosis, for example, in threads of algae that are added to them. Under these conditions, absolutely no water is secreted from fleshy plants, even after an extended period of time when plants have already been damaged, as eventually occurs with any narcosis that lasts too long. One glance at Dubois' figures [7] will in any case immediately indicate to any botanist that in these experiments, the test plants were not merely narcotized by the anaesthetic, but in fact, were partially destroyed. Dubois is no more successful when he draws a parallel between this secretion of water in too strongly narcotized plants, and water secretion in frozen plants, and assumes that narcosis and rigidity from cold are in principle produced by the same causes (withdrawal of the intracellular fluid from the protoplasm). The reduction and final cessation of signs of life as the temperature decreases occur both gradually and continuously. A graph of the intensity of the individual indicators of life would show no irregularities, but gradually decrease with no break in the abscissa as the temperature decreases. Water secretion does not take place until the plants are actually frozen, and in this case, it is related to a disturbance of the normal osmotic properties of the protoplasm. The disappearance of haemoglobin from red blood corpuscles after storage of frozen blood, or after the addition of large amounts of ether, chloroform, and other such substances also results from a disturbance of the normal osmotic properties of the blood corpuscle flow, as in the case with all plant and animal cells subjected to fatal toxicants. The disappearance of haemoglobin from the blood corpuscles is in principle the same process as the disappearance of the coloured cell fluid from plant cells after these have been killed by too high temperatures, freezing, the addition of toxicants, and other external effects. The concentrations of most toxicants that are needed for both of these responses to occur are therefore in many cases, essentially the same. However, like protein, haemoglobin is coagulated by the action of many toxicants, and as a result, is prevented from passing out of the blood corpuscles.

Although Dubois' experiments were incorrectly interpreted, this does not imply that his hypothesis is entirely erroneous. I should like to point out that, if muscles which have been fully narcotized (though not killed) by chloroform, ether, or other narcotics suffer water loss, this loss must, in any case, be much smaller than can be produced by non-specific dehydrating agents (salt and sugar solutions, which possess a greater osmotic pressure than the blood of the test animal) without causing loss of muscle sensitivity. If water loss does play a more significant part in narcosis produced by actual narcotics, then only specific components of the protoplasm can become substantially lower in water, not the entire protoplasm as it is when affected by stronger salt solutions, and other such dehydrating agents. The same thing can also be readily demonstrated for the meristem cells of plants.

2.5 RICHET'S PRINCIPLE

For several years, Richet has presented a concept on various occasions which, although it cannot be exactly considered a theory of narcosis, nevertheless should be considered as an explanation of narcosis, even though it is by no means generally applicable. According to Richet, the lower the solubility of a compound in water, the more potent a narcotic, or generally speaking, a toxicant, the compound is. As we shall see later, this statement applies to non-specific narcotics for the members of a homologous series of compounds, and to a lesser extent, also to other chemically related compounds, although only up to a certain member of the series. Later, we will become acquainted with the correct interpretation of this phenomenon, to the extent that it is valid. Richet believes that a compound that does not dissolve (i.e. that is only slightly soluble in water) is probably therefore toxic to a cell because it does not diffuse evenly into the protoplasm [8]. If, as it seems to me, Richet is trying to say that the protoplasm sets up considerable resistance to the diffusion in water of compounds that do not dissolve much, this is not at all true. Richet is trapped in the same error as Dubois, when the latter designates anaesthetics as dysosmotic. I have already indicated in my critique of Dubois' definition that this description is entirely misleading. All of the non-specific narcotics that are not very soluble in water, and the majority of the toxicants that are not very soluble in water, spread very quickly through the entire protoplasm, and reach the cell fluid in plant cells in a few seconds or minutes (assuming readily permeable cell membranes). As already emphasized, the ability to spread quickly through the entire living protoplasm is, in fact, a condition *sine qua non* for the non-specific narcotics.

2.6 HYPOTHESES BASED UPON THE CHEMICAL COMPOSITION OF THE BRAIN

2.6.1 Chemistry of the nervous system

In the preceding discussion of hypotheses of narcosis, the particular chemical constitution of the nervous system, or rather of the ganglia cells (i.e. the cells first affected by narcotics) has not played any role. However, it is this particular chemical composition of the ganglia cells that is the basis for the hypotheses that remain to be discussed.

It occurred to researchers occupied with the chemical investigation of animal tissue long ago that the nervous system is especially rich in compounds that are soluble in ether and that, among these substances, cholesterol and a 'phosphorus-containing' fat, lecithin, predominated. For some time, the presence of these compounds was known only in the gall and nervous systems. Only later were they discovered in other animal tissues and plants, where they are present in much smaller amounts. According to several hypotheses of narcosis, those components of the nervous system that are soluble in ether are the sites of action of the narcotics (or part of the narcotics). However, the individual hypotheses differ with respect to the assumed form of action. In some cases, moreover, they were developed independently of one another. Before discussing them specifically, it may be appropriate to provide a brief overview of the present state of our knowledge of these components of the nervous system that are soluble in ether, with regard namely to their distribution in the nervous system.

Besides lecithin and cholesterol, protagone and the cerebrines also belong with these compounds. It is unknown, however, whether the latter are formed within the nervous system. In addition to these substances, another substance known as jecorine has been reported to occur. However, there has recently been some dispute regarding whether jecorine represents a unique substance. Of particular importance is the question of how these compounds are distributed in the grey and white material of the nervous system. This is because narcosis results mainly from a change in the grey substance, since only vertebrates possess the nervous medulla. Nevertheless, invertebrates can be just as readily narcotized as vertebrates, and it appears all the more certain that the mechanism of narcosis must be essentially the same in both types of animals since the concentrations of non-specific narcotics sufficient for complete narcosis of entomostraca (this does not apply to all invertebrates, however) and mammals are almost exactly the same (at least this is the case for ethyl ether, chloroform, and other narcotics), as will be demonstrated later.

Petrowski [9] was the first to investigate in more detail the qualitative and quantitative distribution of ether soluble constituents of the nervous system between the grey and white matter. Despite criticism of his research, Petrowski's work remains the most important on this subject. Petrowski attempted to analyse the grey and white matter of the brain which had been separated as much as possible from each other. It is impossible, of course, to obtain grey matter that is completely free of white matter from mammals after birth. Nevertheless, grey matter which has been purified as much as possible exhibits an entirely different chemical constitution from white matter. The white matter is known to consist of axial cylinders (whose quantitative chemical composition is probably equivalent to that of the grey matter) and the nervous medulla, to which the white matter owes its characteristic physical and chemical properties. Table 2.1. shows Petrowski's reported composition of dried grey and white matter isolated from the ox brain.

In analysing these constituents, Petrowski assumed that lecithin is the only constituent of the nervous system that contains phosphorus soluble in alcohol and ether. Like Hoppe-Seyler, he considered protagone to be a mixture of lecithin and cerebrine. Subsequent research by Gamgee and Blackenhorn [10], and Baumstark and Ruppel [11] and Kossel and Freytag [12] has of course shown that protagone is a discrete compound, and not a mechanical mixture of lecithin and cerebrine. In the normal study of cerebrines, the protagone in cerebrine and other compounds is released. Petrowski obtained only 0.53% cerebrine from dried grey matter, which in any case, was probably derived from small quantities of associated white matter. Thus, it follows that protagone, if it occurs at all, appears only as traces in the grey matter, but occurs in abundance in the white matter.

It seems all the more probable that the grey matter of the brain contains no protagone or cerebrine, when a comparison is made of the adult brain with that of the embryonic brain prior to the appearance of the nervous medulla. The entire brain of these embryos consists almost exclusively of grey matter. Raske [13] found the brain in ox embryos contains neither protagone nor cerebrine. Raske's figures for the percentages of lecithin and cholesterol in dried brains are given in Table 2.2.

Table 2.1 The composition of dried grey and white matter isolated from the ox brain

	Protein (%)	*Lecithin (%)*	*Cholesterol and fat (%)*	*Cerebrines (%)*	*Neuro-keratins and other organic compounds (%)*	*Salts (%)*
Grey matter	55.37	17.24	18.86	0.53	6.71	1.45
White matter	24.72	9.90	51.91	9.55	3.34	0.57

Table 2.2 Lecithin and cholesterol in dried brains

	Ox embryo (62 cm long) (%)	*Ox embryo (68 cm long) (%)*
Lecithin	6.63	3.49
Cholesterol	18.32	21.32

The cholesterol content of these embryonic brains agrees approximately with the amounts found by Petrowski for the grey matter of the adult brain. On the other hand, according to these figures, the lecithin content of the embryonic brains would be much lower than in the grey matter of the fully developed brains. Raske assumes that the lecithin content of the ganglia cells and the axial cylinders increases significantly during the later development of the brain. The great discrepancy between the amounts of lecithin found in the two embryos is remarkable and it would be very desirable to initiate as soon as possible a larger series of studies of the lecithin and cholesterol levels in the brains of embryos of different ages. The use of Raske's results could be objectionable since protagone might not be formed in the ganglia cells and the axial cylinders until the final stages of brain development. There are, however, other grounds for rejecting the presence of protagone even in the fully developed grey matter of the brain and in the axial cylinders. According to Wlassak [14], the Weigert method of staining the nervous medulla depends mainly upon its protagone content. Since the ganglia cells and the axial cylinders do not become stained when the Weigert method is used, it can be concluded that they contain no protagone or only traces of it.

It should be mentioned that according to Baumstark, only a part of the cholesterol in the nervous system exists in the free form, while another part is present in the form of an ester. However, since Baumstark did not investigate the grey and white matter separately, it cannot be demonstrated with certainty at the moment whether the cholesterol ester occurs in both the grey and white matter or, as is more probable, only in the white. The true fats found in the brain probably belong exclusively, or almost exclusively, to the connective tissue.

After weighing the results of these various investigations, it can be concluded that lecithin and cholesterol are at the moment the only known constituents of the grey matter of the brain that are soluble in ether. The quantitative figures of Petrowski, as far as they apply only to the grey matter, can be considered for the time being to be more or less correct, though further studies of this subject would appear to be very desirable. Thus, lecithin and cholesterol occur in about the same amounts in the grey matter, consisting of about 18% each in the dried material, and slightly more than 3% of the fresh material. Since the narcotic content of the blood plasma remains constant, the ratios in the nervous medulla do not require further consideration. On the other hand, if the test animal is given a certain weight of the narcotic per kilogram of its body weight, then, of course, the concentration of the narcotic in the blood plasma can be altered by the accumulation of the narcotic in the nervous medulla in the same way that it can be changed by its accumulation in the fatty tissue.

2.6.2 Hypothesis of Bibra and Harless

An introductory overview of the chemical constitution of the nervous system and in particular of the grey matter of the brain has been provided to the extent that it seems important for an understanding and assessment of hypotheses on narcosis. A critical discussion of those hypotheses relating solely to the action of the non-specific narcotics follows. By 1847, immediately following the discovery of the anaesthetizing properties of ether and chloroform, Bibra and Harless had already attributed the capacity of various anaesthetics to dissolve fats as being the determining factor in their action. These authors, who, incidentally, based this conclusion on a very limited amount of data, understood the anaesthetics to act in such a manner that they extract the 'brain fats' from the ganglia cells. They postulated that following narcosis with ethyl ether, the fats extractable by ether undergo a loss from the brain with an increase in the liver. For Bibra and Harless to reach this conclusion from their studies, it could only have been the result of a chance occurrence in the analytical results. First of all, the fundamental phenomenon of narcosis induced by non-specific narcotics, namely that narcosis immediately fades as soon as the concentration of the narcotic in the blood plasma falls below a certain level, would remain entirely unexplained if such an extraction out of the ganglia cells really did take place. Secondly, the 'brain fats' are essentially just as insoluble in very diluted aqueous narcotic solutions which can induce narcosis as in pure water.

A few concrete examples are sufficient to support these objections. For example, in order to produce complete ether narcosis, the concentration of the ethyl ether in the blood plasma must be almost exactly 0.25% by weight, and when this concentration is reached, narcosis occurs immediately without subsequently becoming much deeper. If narcosis really depended upon the extraction of 'brain fats' from the brain cells, first of all, the narcosis would have to deepen with time, since this extraction would be rather lengthy, and secondly, it must logically be assumed that once narcosis passed, for example, if the ether concentration in the blood fell to 0.17%, there would be at least a partial return of the brain fats into the ganglia cells. It would be very difficult, however, to find an explanation for such a process. Furthermore, it is highly probable that the mechanism of ether or chloroform narcosis, for example, remains substantially the same in the ganglia cells, the ciliary cells, and in plant cells as well. An algal thread or a ciliary cell placed in a solution of ether or chloroform would release these fatty compounds into the surrounding solution. In fact, if the volume of the solution were sufficiently large, the extraction would be complete. The same result would also have to occur in fish and other gill-breathing animals that are placed in a narcotizing solution. As soon as these algae, ciliary cells, or tadpoles are transferred to pure water, the narcosis fades immediately, even though the possibility of a return of the fatty compounds into these cells is excluded. Therefore, the extraction process described cannot take place when narcosis occurs. As we shall see later, chloroform produces complete narcosis at a blood plasma concentration of only 1 : 6000 by weight; for phenanthrene, whose mechanism of action is certainly quite similar to that of chloroform, this occurs even at a concentration of 1 : 1 500 000. It is perfectly clear that such diluted aqueous solutions of the narcotics cannot exert a dissolving effect in the sense of an extraction of the fatty substances of the ganglia cells. Under these circumstances, it is not surprising that the hypothesis of Bibra and Harless was soon forgotten.

2.6.3 Contribution of Hermann

After Hermann had identified the lecithin content of red blood corpuscles, he also raised the question [15] whether perhaps the lecithin, cholesterol and fats of the ganglia cells and the red corpuscles might not be the common site for the action of anaesthetizing compounds. It was known that a number of these compounds produce a 'dissolving' of the red blood corpuscles in addition to having a narcotizing effect on the brain (although only at much higher concentrations). This is known to be the case, for example, with chloroform (Boettcher), ethyl ether (Wittich and Hermann), and ethyl alcohol.

Hermann never discussed in great detail his understanding of the action of anaesthetics on lecithin, cholesterol, and related compounds. Based upon his statement that many anaesthetics are good solvents for these compounds, it may perhaps be inferred that Hermann's concept of the action of anaesthetics was similar to that of Bibra and Harless.

It has already been emphasized that the 'dissolving' of the red corpuscles corresponds substantially to the processes that occur in all other plant and animal cells following their death. It is connected with a change in their osmotic properties following cell death. It is not certain at this time whether in the particular case of fatal cell intoxication by non-specific narcotics that the change in the osmotic properties of the cell can be attributed to a direct effect of these narcotics on the lecithin and cholesterol-related components of the cell. This is not unlikely. However, it must be emphasized that, if this is true of the red corpuscles, it must be no less true for other plant and animal cells. As I indicated earlier, it seems to me that the theory that non-specific narcotics act on the constituents of the cell, including lecithin and cholesterol, when narcosis of the neurons or dissolving of the red corpuscles occurs, is much less certain than a second theory; namely, that essentially the same mechanism produces narcosis in the neurons and other cell tissues (either plant or animal) that is caused by non-specific narcotics. It is also very probable that the non-specific narcotics exert their effect principally on the cholesterol and lecithin-related constituents of the cells, but not in the way visualized by Bibra and Harless. These compounds probably change the normal physical state of these cell constituents without causing them to be removed from the cells.

2.7 THEORY OF H. MEYER AND THE AUTHOR ON NARCOSIS INDUCED BY NON-SPECIFIC NARCOTICS

It is precisely in these last named points, aside from their more general application, that the theory of narcosis by non-specific narcotics arrived at independently by Meyer and the author differs from earlier hypotheses of Bibra and Harless, and of Hermann, which in any case, had been almost completely forgotten. Instead of starting out with the solubility of the cholesterols, lecithins, and related materials in certain anaesthetics, Meyer and this author chose as a starting point of their theory the reverse idea, the solubility of non-specific narcotics in the cholesterol and lecithin-related compounds in the cell. They made the assumption that narcosis results from the modification (caused precisely by the absorption of foreign compounds) of the normal physical state of the lecithin and cholesterol-related compounds in the cell.

It can easily be shown that in many cases non-specific narcotics have no chemical effect on these compounds and that it may therefore simply be a question of a change in the physical state of the lecithins and related compounds. Thus, the amount of foreign compound absorbed will primarily determine its degree of effect. Nevertheless, the possibility is not excluded that the qualitative nature of the compound has some importance. There may also be some doubt as to whether it depends mainly on the number of molecules of narcotic that are absorbed by a given amount of brain lipoids (lecithin and cholesterol-related constituents of the cells), or rather on the volume of narcotic absorbed. In both cases, however, the action of the narcotic will depend mainly on the partition coefficient of the narcotic between water, on the one hand, and the brain lipoids, on the other. This theoretical conclusion can be confirmed by experiment, and has been verified to such a large extent in the experiments of both Meyer and myself that this theory possesses a very high degree of probability.

If we already knew the exact quantitative composition of the grey matter in the nervous system, or of its components that are soluble in ether, then, as I will demonstrate later, it would be possible in many cases to determine exactly the partition coefficients between water and the brain lipoids, and this would undoubtedly be in principle the most accurate. Meanwhile, however, lecithin, which forms a considerable part of this mixture, is at present difficult to obtain in fairly large quantities; and in addition the experiments carried out so far on the quantitative composition of the grey matter are scarcely adequate for more than mere orientation. Therefore, any such direct experiments will have to be left for the future.

Both Meyer and this author have mostly been satisfied for the present in determining the partition coefficients of non-specific narcotics between water and olive oil. It generally appears that, if the partition coefficient of a compound between water and olive oil favours the latter, it will also favour the higher alcohols, ethyl ether, and similar solvents, if these are used instead of olive oil as the second solvent. Of course, it must be emphasized that this only applies on the whole. The partition coefficient of non-specific narcotics between water and olive oil will hardly ever be exactly the same as that between water and the brain lipoid mixture in the grey nervous matter, but will only be as close as the partition coefficient of the same narcotic between water and ethyl ether. Even the qualitative dissolving capacity of the brain lipoids does not correspond entirely with that of olive oil; the commercial salts of the basic aniline dyes are completely, or almost completely, insoluble in olive oil, whereas they readily dissolve in both cholesterol (in the melted or dissolved state) and lecithin (see Overton, E. (1900) 'Studien über die Aufnahme der Anilinfarben durch die lebende Zelle.' *Jahr. wiss. Bot.*, **34**, 669–701).

My experience up to now has demonstrated that the dissolving capacity of melted cholesterol corresponds exactly with that of the higher saturated alcohols, e.g. ethal ($C_{16}H_{33}OH$) or ceryl alcohol ($C_{27}H_{55}OH$), and the dissolving capacity of lecithin, despite its chemical composition, also corresponds more closely with that of the higher alcohols than it does with the dissolving capacity of the fatty oils. Unfortunately, this did not become known to me until almost two years ago, at a time when my experiments on narcosis had reached a temporary halt.

In any case, the higher alcohols from $C_9H_{19}OH$ up to $C_{15}H_{31}OH$ are very difficult to obtain, and even ethal and ceryl alcohol are so expensive that to determine the partition coefficients of a larger number of narcotics between these and water would be a very costly undertaking, at any rate, if there were a desire to determine the partition coefficients by both physical and chemical methods. In addition, this measurement could be conducted only at a temperature that greatly exceeds the body temperature of mammals, and deviates even further from those suitable for narcosis with gill-breathing animals. This is because ethal has a melting point of 49.5°C, and ceryl alcohol does not melt below 79°C, and also because partition coefficients may exhibit a large or small variability with temperature. Nevertheless, the properties of these two alcohols are much more favourable than that of cholesterol, which does not melt below 147°C.

Since the precise quantitative composition of the lecithin–cholesterol mixture of the grey nervous matter is unknown, and the partition coefficients of the various narcotics between this mixture and water cannot be measured, it must suffice to use olive oil or a suitable solvent in experiments on partition coefficients. It would appear from prior studies that it is far more important in testing a new theory of narcosis, to determine the approximate partition coefficients of as large a number of non-specific narcotics as possible, rather than determine with particular exactness the partition coefficients of a smaller number.

REFERENCES

1. Durham, A.E. (1860) The physiology of sleep. *Guy's Hospital Reports*, third series, **6**, 149.
2. Hammond, W.A. (1866) *On Wakefulness*, Philadelphia, (quoted according to Claude Bernard).
3. Bedford-Brown (1860) *American Journal of Medical Science*, October (quoted according to Claude Bernard).
4. Bernard, C. (1875) *Leçons sur les Anesthésiques et sur l'Asphyxie*, Librairie J.-B., Ballière et Fils, Paris, pp. 117–18.
5. Bernard, C. (1875) *Leçons sur les Anesthésiques et sur L'Asphyxie*, Librairie J.-B., Ballière et Fils, Paris, p. 153.
6. Dubois, R. (1894) *Anesthésie Physiologique*, p. 15. This hypothesis is also clearly cited earlier by him (1885), *Soc. de Biologie*, October 24, (1888), October 27, and various published sources.
7. Dubois, R. (1894) *Anesthésie Physiologique*, p. 15, Fig. 2.

8. Richet, C.R. (1895) *Dictionnaire de Physiologie*, F. Alan, Paris, Vol. I, article on alcohol.
9. Petrowski, D. (1873) Zusammensetzung der grauen und der weissen substanz des Gehirns, Pflüger, *Arch. ges. Physiologie*, **7**, p. 367–70.
10. Gamgee, A. and Blackenhorn, E. (1880) Ueber Protagon, *Journal of Physiology*, Cambridge and London, **2**, p. 113–31. Gamgee, A., *Physiol. Chem.*, **1**, p. 427.
11. Baumstark, F. (1885) Ueber eine neue Methode, das Gehirn chemisch zu erfreschen, und deren bisherige Ergebnisse, *Ztscht. Physiol. Chem.*, **9**, 145–210.
12. Kossel, A.C. and Freytag, F. (1893) Ueber einige Bestandtheile des nervenmarks und ihre verbreitung in den Geweben des Thierkörpers, *Ztschr. Physiol. Chem.*, **17**, 431–56.
13. Raske, K. (1886) Zur chemischen Kentniss des Embryo, *Ztschr. Physiol. Chem.*, **10**, 336–45.
14. Wlassak, R. (1898) Die Herkunft des Myelins, *Arch. Entwicklungsmechanik der Organismen*, **6** (4), 453–94 (particularly pp. 458–61).
15. Hermann, L. (1866) Uber die wirkungsweise einer Gruppe von Giften, *Arch. Anat., Physiol.* Il. Wiss. Med., p. 27, (1874) *Lehrbuch der Experimentellen Toxicologie*, August Hirschwald, Berlin.

3

Lipoid theory of narcosis and partition coefficients

3.1 THEORY OF PARTITION COEFFICIENTS

Because of the great importance of partition coefficients of narcotics between water and organic solvents in connection with the new theory of narcosis, it would not be inappropriate to comment on the theory of partitioning of a compound between two solvents that are immiscible with one another, and on the methods that can be used to measure partition coefficients (*cf* also Nernst (1898) *Theoretische Chemie,* 2nd edn, pp. 454–6; and (1889) *Zeitschr. physik. Chem.*, **4**, 150, and (1890) **6**, 16). It is assumed that the solvents are essentially insoluble in one another, as is actually the case, for example, for olive oil and water, xylene and water, and petroleum ether and water.

We assume first of all that the compound whose partition coefficient we wish to measure is soluble to only a relatively small degree in the two chosen solvents, e.g. that the solubility of the compound in the stronger solvent at the corresponding temperature is at the most 8–10%. Under these conditions, the compound is distributed between the two solvents at temperature T, regardless of its concentration, in such as fashion that, after equilibrium has been reached, the quantities of compounds contained in the same volume of solvents are always proportional to their respective solubilities in the two solvents. If, for example, at 20°C, 1 gram of the compound is soluble in 1 litre of water and at the same temperature 10 grams of the compound are soluble in 1 litre of xylene, petroleum ether or olive oil, then the compound will always distribute itself between the water and the xylene (or petroleum ether or oil) at 20°C in such a way that the xylene contains ten times as much of the compound as the same volume of water, regardless of whether it is a concentrated or more dilute solution of the compound. The partition coefficient of the compound between water and xylene at any concentration is therefore 10/1 = 10, if the amount of the compound contained in a unit volume of the aqueous solution (after partitioning) is set to 1.

The compound, of course, must be present in the two solvents in the same molecular state, or one that should at least not change noticeably with dilution. For this principle to be valid, the compound may not, for example, be present in the organic solvent solely in the non-dissociated form, and in the water partly in the form of ions. Similarly, the compound may not be present in one solvent in the monomeric form and in the other solvent partly as dimers, at least in so far as the relative proportions of monomers to dimers changes with dilution in the second solvent (which will generally be the case). Since all non-specific narcotics are non-electrolytes, the first qualification requires no further consideration. Furthermore, all non-specific narcotics which have no greater than 10% solubility in either water or the organic solvent will presumably exist only as monomers in solution. If a small percentage of dimeric molecules are present in the organic phase, their number will be too small to affect the partition results significantly.

The constancy of the partition coefficients for the most varied concentrations of compounds that fulfil the above criteria is an experimentally proven fact. This, however, can also be inferred theoretically on the basis of Henry's Law of the absorption of gases in liquids if this is also applied to vapours. This extension of the law to vapours is justified and could already have been predicted a priori with almost complete certainty since there is no significant difference between gases and vapours. It can, however, also be easily verified experimentally for the vapours of chloroform, ethyl ether, ethyl bromide and many other compounds.

According to Henry's Law, the amount of gas absorbed by a given amount of liquid is directly proportional to the pressure of the gas. However, the amount of gas absorbed depends, aside from the chemical nature of the gas, on both the nature of the absorbing liquid and the temperature. If a unit volume is used as a measure of the amount of absorbing liquid, then the law can be formulated as follows: when a gas is absorbed by a liquid, after equilibrium occurs, the concentration of the gas in the liquid is always proportional to the gas pressure. Therefore, if a liquid being contained in an enclosed space that is not completely filled by the liquid dissolves more of the gas at the beginning of the experiment than would correspond to the gas pressure in the unfilled space, the gas will be released from the liquid into the vapour volume. However, if at the beginning of the experiment the liquid dissolves less of the gas than would correspond to the gas pressure, the gas will be absorbed from the gas volume until the conditions of Henry's Law are satisfied.

At room temperature, all non-specific narcotics have a definite, but frequently very low, vapour or sublimation pressure. (Whether this applies to all chemical compounds is uncertain; in the case of proteins and other compounds with very high molecular weights one may doubt it and may prefer to think in terms of a decomposition pressure.) In most cases, this can be recognized by their odour at room temperature, as in the case of phenanthrene, which has a boiling point of 340°C. Furthermore, the partial vapour pressure of a narcotic A in a saturated aqueous solution at temperature T must be $p_{a,t}$ if the vapour pressure of the same narcotic at the same temperature in its pure undissolved state is $p_{a,t}$, since otherwise the solution would have to be either unsaturated or supersaturated. If the concentration of the narcotic in this solution is C, then, if we extend Henry's Law to vapours, the partial vapour pressure of the narcotic in a solution of a concentration of C at the same temperature T must be $p_{a,t}$ and in a solution of a concentration $\frac{1}{2}C$, it must be $\frac{1}{2}p_{a,t}$ and at a concentration $1/m\,C$, $1/m\ p_{a,t}$. The same relationships apply for the solutions of the narcotic in a second solvent, e.g. olive oil, in which the saturated solution at the same temperature T is at concentration C'. Since the partial vapour pressure of the narcotic in the saturated solution is also $p_{a,t}$, it follows that the partition coefficient of the narcotic between equal volumes of the two solvents must be constant and must correspond to the expression C'/C, where C and C' are the concentrations at saturation of the narcotic in the two corresponding solvents.

Partition coefficients are more or less variable with temperature, due to the fact that the solubility of a compound in different solvents does not increase or decrease in equal proportion within a finite temperature change. As is the case with gases, the ratio of the partition coefficients in different solvents varies with temperature. Nevertheless, by analogy to the behaviour of gas absorption in different solvents, the partition coefficients of compounds that are liquid or solid at room temperature usually vary only slightly with small changes in temperature, i.e. no greater than 10–20°C.

Up to now, we have dealt only with the simpler case in which a compound is only moderately soluble in each of the two solvents at the relevant temperature. If the compound is highly soluble in one or both of the solvents, or miscible with one in all proportions, as is the case with ethyl ether, chloroform and many other narcotics in olive oil, xylene, and other solvents, then the ratios become much less straightforward in that the partition coefficients for the concentrated solutions change with concentration. A similar phenomenon also exists with the solubility of gases, in that a gas readily absorbed by a certain quantity of liquid follows

Henry's Law only at very low gas pressure, but not entirely at higher pressures. Nevertheless, even in these cases, the solubility of a compound in two solvents is a reasonable criterion for estimation of partition coefficients between the two solvents. If, for example, a compound is only 1% soluble in water, but is miscible with olive oil in all proportions, it may be confidently assumed that, after partitioning of the compound between water and olive oil, at every concentration there will be at least 20 times more of the compound in the oil than in the same volume of aqueous solution. Thus, the partition coefficient will be greater than 20. This relationship, of course, applies to non-electrolytes or weak electrolytes, a condition that is always the case for non-specific narcotics.

3.2 METHODS FOR MEASURING PARTITION COEFFICIENTS

I have discussed these ratios in some detail because in the experimental section (Part Two) of this study I have frequently provided only the approximate levels of solubility of the various narcotics in water and olive oil, without determining their partition coefficients directly. The methods for directly determining partition coefficients for non-saturated solutions of narcotics can be divided into chemical, physical and physiological methods. Only the latter two methods will be discussed here, since they were always adequate for the sake of this study. The physical methods for measuring partition coefficients vary according to the particular physical properties of the narcotic under study.

3.2.1 Physical methods

Non-volatile narcotics
Stable narcotics with sufficient water solubility, but which are not appreciably volatile at room temperature, can be measured as follows: an aqueous solution of the compound is prepared at a known concentration, and a carefully measured volume of a second solvent (which should not be significantly soluble in the first) is added to a certain volume of this solution. The two solvents are shaken vigorously for some time, after which the mixture is left until complete separation has occurred. Subsequently, the concentration of the narcotic in the aqueous solution is measured a second time by evaporating or desiccating (depending upon the volatility of the narcotic at 100°C and its degree of solubility in water) the water from a measured quantity of the aqueous solution, and weighing the solid residue. If equal volumes of the aqueous solution and the second solution are shaken in a partitioning experiment, the partition

coefficient is equal to the difference in aqueous concentration before and after being shaken with the second solvent, divided by the concentration of the aqueous solution after shaking. For example, if the concentration of the narcotic in the aqueous solution was 2% before shaking, and only 0.2% after being shaken with an equal volume of the second solution, then the partition coefficient is:

$$(\text{Solvent 2}) / (\text{Water}) = (20 - 2) / 2 = 9.$$

If the narcotic is much more readily soluble in the second solvent than in water, it is desirable to shake one volume of this solution with 10, 50, or even 100 volumes of the aqueous solution. To calculate the partition coefficient, the value obtained by dividing the concentration difference of the narcotic in the aqueous solution before and after shaking must be multiplied by 10, 50 or 100. It must be noted, however, that this method is not reliable for compounds such as camphor and thymol since, despite their rather high boiling points, too much of these narcotics is lost by evaporation, unless the compound is not adsorbed by the dehydrating agent within the desiccator.

Volatile narcotics

In dealing with narcotics that are volatile at room temperature, the simplest method for measuring partition coefficients is volumetrically, as long as the narcotic is fairly soluble in both solvents. For example, let 5–10 volumes of the narcotic be shaken with 50 volumes of each of the two solvents. Following separation of the two solvents, the partition coefficient is calculated by dividing the volume increase of the organic solvent by the volume increase of the aqueous solution. If the narcotic is much more soluble in the organic phase than in water, it is convenient here to shake a certain volume of the narcotic with, for example, 90 volumes of water and 10 volumes of the organic solution, and apply the corresponding modifications in the partition coefficient calculation. The volume of the organic solvent, however, must always be much greater than that of the pure narcotic, since otherwise, the partition coefficient will apply only to this concentration, and not to more dilute solutions. (The values provided for the partition coefficients in the experimental section (Part Three) are not always free of this error. The concentrations of narcotics required for narcosis in plant cells are frequently rather close to their saturation values in water.) If the narcotic is considerably more soluble in water than in the organic solvent, then, of course, one should use a larger volume of the organic solvent than water in the partition coefficient experiments. Not infrequently, when water and olive oil are used as solvents for the narcotic, a tough emulsion forms at the boundary between the solutions that can remain for weeks. If this transition phase is not very thick, it only lowers the accuracy of the result. In other cases, however, it may render impossible any accurate determination of the partition coefficient. The separation of the two liquids can often be accelerated with the aid of a centrifuge, which also reduces the emulsion phase, should one form. Finally, it should be noted that it is desirable to perform the experiments twice. In the first case the compound is initially dissolved in water and shaken with oil; in the second case it is initially dissolved in oil and shaken with water. If equilibrium is not reached, this would be obvious from an inequality in the two measurements.

Liquid narcotics with high vapour pressure and slight water solubility

For liquid narcotics having a high vapour pressure at room temperature with only slight water solubility, partition coefficients between water and oil can be determined from their corresponding partial pressures in these two phases. The procedure consists first of determining the partial vapour pressure of the narcotic at temperature T for a solution of the narcotic in oil of a certain concentration. In a second experiment, the concentration of the aqueous solution of the narcotic having the same partial vapour pressure of the narcotic at temperature T is determined by experiment or, more simply, by calculation. The partition coefficient is then equal to the concentration of the narcotic in oil, divided by the corresponding concentration in aqueous solution. In order to calculate the latter, it is necessary to know only the vapour pressure of the pure, undissolved narcotic and its solubility in water at the relevant temperature, for which data are available in many cases. The partial vapour pressure of the narcotic in the saturated aqueous solution is approximately equal to the vapour pressure of the undissolved narcotic at the same temperature. Both vapour pressures would be exactly equal if water were completely insoluble in the narcotic. However, the vapour pressure of the narcotic in aqueous solution is somewhat lower. Nevertheless, it can simply be considered equivalent when great accuracy is not needed in the partition coefficient determinations. Thus, it follows from Henry's Law that the desired concentration of the narcotic in aqueous solution corresponds to the concentration of the saturated aqueous solution of the narcotic in the same way that the vapour pressure of the narcotic in oil solution corresponds to the vapour pressure of the narcotic at the same temperature in its undissolved state. This method is the best means of determining the partition coefficients of chloroform, ethyl bromide, ethyl ether and similar substances between water and oil if there is a preference for physical methods of measurement. It should, incidentally, be kept in mind that the above narcotics are miscible with oil in all proportions. Therefore, the partition coefficients will vary with concentration for concentrated solutions. Thus, partition coefficients for these solutions should be measured at about the same concentration in the aqueous phase after being shaken with oil, as is required to induce narcosis. This approach should be employed so long as the relative degree of experimental error would not then be increased. Usually, it is more practical to measure the vapour pressure of the narcotics in oil solutions at somewhat higher concentrations.

3.2.2 Physiological methods

The partition coefficient of a non-specific narcotic between water and an organic solvent can be measured by a physiological method if the saturated solution of the relevant organic solvent in water produces no noticeable effect on the organisms used in the experiment. This condition is satisfied with neutral olive oil, but not, for example, with xylene, petroleum ether and most other organic solvents. Despite the extraordinarily low solubility of many such solvents in water, in saturated aqueous solutions they have a narcotic or even an immediately fatal effect. When this method is used, certain aquatic animals that breathe through the gills or through the skin, or even plant cells, are useful as indicators for the concentration of the narcotic in the aqueous solution after it has been shaken with the olive oil. The most suitable organisms are young tadpoles 9–14 mm long and various entomostraca (varieties of daphnia, bosmina, cyclops, and related organisms).

In order to measure the partition of a non-specific narcotic between water and oil by means of this physiological method, the concentration of the narcotic in aqueous solution that is just sufficient to completely narcotize the indicator organisms must be determined. Not all aquatic animals are narcotized at the same concentration. Non-specific narcotics that are much more soluble in oil than in water affect the *Naiadaceae* (naias, chaetogaster, aeolosoma) or nephelis (belonging to the *Hirudinea*) at about double the concentration of the narcotic that is sufficient to completely narcotize tadpoles and entomostraca. Even higher concentrations are required to narcotize infusoria, plant cells, and related organisms.

Let us assume that the original concentration of the narcotic in aqueous solution is a, and that concentration b is required to narcotize tadpoles or entomostraca, while concentration $2b$ is needed to narcotize varieties of the genus naias, chaetogaster, and related organisms. If after shaking equal volumes of the original aqueous solution and olive oil, the concentration of the narcotic remaining in the aqueous solution is sufficient to narcotize tadpoles but not naiads, this indicates that the concentration of the aqueous solution after being shaken with oil lies between the values *b* and 2*b*. This aqueous solution is separated from that of the narcotic in olive oil, and its concentration is measured. The solution is subsequently diluted with water until narcosis no longer occurs with tadpoles or, better, until existing narcosis partially disappears. By noting the quantity of water added, a determination can be made immediately of the original aqueous concentration of the narcotic following shaking with oil. If this concentration is divided by the difference in the concentration of the aqueous solution before and after being shaken with oil, this will provide the partition coefficient, or rather its reciprocal value.

Although it is not essential, it is desirable to simultaneously employ both tadpoles and naiads as indicators. In addition, it should be noted that the length of time required for narcosis to occur without causing death can provide a valuable criterion for estimating if a solution possesses a concentration just sufficient to induce narcosis, or far exceeds it. A solution twice as concentrated as the threshold value is usually fatal within an hour, for example, while one exceeding it by a half, is fatal after about three hours. There are, of course, significant differences between individual narcotics in this respect, and these need to be known ahead of time. Those narcotics much more soluble in olive oil than in water, should not be shaken with an equal volume of olive oil, but with only 1/10, 1/50, 1/200, and even smaller proportions, which will then require a corresponding change in the calculation of the partition coefficient.

There are many great advantages of using this physiological method of determining partition coefficients between water and olive oil, and it can be varied in many ways depending upon the situation. Thus, it can be used for all non-specific narcotics, even those soluble in more than 100 000 parts of water. These require only small amounts of the narcotic, since just a few millilitres of the solution are needed for the narcosis test, so long as only small tadpoles or entomostraca measuring just a few centimetres are used as indicators, except in the case of very dilute solutions of the narcotic. Under these conditions, the rule applies that the more dilute a solution required for narcosis the greater the quantities of solution needed, so that as the narcotic is absorbed by the test organism no significant change in the concentration of the surrounding solution takes place. As one advantage, this method permits partition coefficients to be determined specifically at the concentrations of narcotics that are very close to the narcotic threshold concentration. It has already been emphasized that partition coefficients vary with temperature when the narcotic is very soluble in olive oil (or in water) or is miscible with these substances in any proportion. In order to make these particular determinations, it is only necessary to modify this method. Instead of diluting the aqueous solution of the narcotic after it separates from the oil solution with pure water until no further narcosis occurs, a specific volume of the original narcotic aqueous solution is shaken with various quantities of oil, and the volume of oil noted that removes enough of the narcotic from the aqueous solution so that the concentration is just sufficient to produce narcosis in tadpoles or entomostraca. In addition, this modification of the method must always be made when the aqueous or olive oil solutions of the narcotic cannot be separated completely; i.e. when a thin emulsion of oil remains in the water.

Unfortunately, I came upon this physiological method of measuring partition coefficients too late in my studies to be able to use it more extensively. Otherwise, I would have applied it in many ways due to its simplicity and elegance.

The use of appropriate organisms as indicators of specific quantitative proportions may, in general, become an aid to physiological research of the most varied kind, especially in cases in which either only very small quantities of the test solution are available, or in which chemical and physical methods are poorly developed or too cumbersome. (This method was originally devised by the author in order to measure the quantitative behaviour of narcotics on transfer to the osmotic pressure of the root sap, and it has been used by him in many other ways.) It cannot be emphasized too often that in physiological research, in those cases in which individual differences in the test subjects, or special circumstances that cannot be accounted for cause quantitative deviations in the individual test results, it is always much more important to have a method of study that permits a large amount of data of reasonable accuracy to be obtained within a short time, than one that may provide somewhat more precise results, but due to its time-consuming nature, permits only a few tests. After discovering the reasons for discrepancies in individual experiments, this greater degree of accuracy is worthless, even for later research, unless by chance (which is usually not the case) further information regarding these causes is available in the test reports. It is normal that when living organisms are used as indicators of certain quantitative relationships that the experiments must be conducted carefully, and various control tests must be performed.

3.3 MEASUREMENT OF PARTITION COEFFICIENTS BETWEEN WATER AND CEREBRAL LIPOIDS

It was indicated previously that if lecithin, or a mixture of lecithin and cholesterol is added to an aqueous solution of a narcotic that is much more soluble in olive oil, ethyl ether, or other such substance than in water, a significant portion of the narcotic will be removed from the aqueous solution when this portion is absorbed by the lecithin–cholesterol mixture. At the same time, it was stated that it is possible, at least in principle, to determine the partition coefficients of various narcotics between an aqueous solution and a lecithin–cholesterol mixture. The first of these statements can easily be proven by using small tadpoles and entomostraca as indicators of the concentrations of the narcotic in the aqueous solution. The actual determination of the partition coefficients is equally certain in many cases with the aid of this method, although it is rather time-consuming, and requires large quantities of lecithin. In addition, the gradual decomposition of the lecithin as is usual in this swelled state presents a problem. Nevertheless, I am unaware of any other current method by which to achieve this goal. Once the percent composition of the brain lipoids mixture is more precisely known, it will be worthwhile to attempt to determine partition coefficients of narcotics between the mixture in these brain lipoids and water by this method.

No study has yet been performed regarding whether the above lecithin–cholesterol releases any of its accumulated water when it absorbs a narcotic from dilute aqueous solution and this cannot be predicted based upon theoretical considerations. (Baumstark's observations of the excretion of water when a brain is placed in pure ether are not relevant to this question, since the use of liquid ether instead of a diluted aqueous ether solution represents an entirely different condition.) If this water accumulation is considered to be a solution of water in the lecithin–cholesterol mixture, then the loss of water would not seem necessary when only a fairly small amount of narcotic is absorbed, since in general, the dissolving of compound *b* in the solution of compound *a* does not lower the solubility of compound *a*. Nevertheless, it is uncertain whether the water uptake by lecithin or lecithin–cholesterol actually represents a solution of water in these substances. Even if this were the case, the quantity of water would be so substantial that the laws governing dilute solutions can hardly be applied to these proportions.

3.4 GENERAL FOUNDATION OF THE LIPOID THEORY OF NARCOSIS

Following the above detailed discussion of the theory of partition coefficients and methods for their determination, we will return to the new theory of narcosis produced by non-specific narcotics.

Meyer [1] reached his theory of narcosis in two ways: first, he realized the impossibility of accounting for narcosis by non-specific narcotics based upon a chemical effect of the narcotic on the tissue of the nervous system, since many of the strongest non-specific narcotics represent the most inactive compounds. Secondly, as a result of discovering that the most varied chemical compounds, so long as they are soluble in ether, have a narcotic effect, and in particular, after learning from testing many of the more important narcotics, that their relative narcotic strengths are in the same order as the relative size of their oil/water partition coefficients. Meyer summarizes his theory in the following statements:

1. All substances that are predominantly chemically unreactive and that are soluble in fat and fatty substances must have a narcotic effect on living protoplasm to the extent to which they are able to be distributed in it.
2. Their action must occur first and strongest within those cells with a chemical composition consisting mainly of fatty substances and which are particularly good bearers of cell functions: primarily, therefore, the nerve cells.
3. The relative strength of action of these narcotics must be dependent upon both their mechanical affinity to fatty substances, and to other cell constituents; i.e. mainly water for the latter. Therefore, it must depend upon the partition coefficient, which determines its distribution in a mixture of water and fatty substances.

Meyer is, of course, aware that the majority of these non-electrolyte compounds that are soluble in fats produce side effects. In certain cases, which we shall encounter later, these side effects may be sufficiently strong that they completely mask the narcotic effect, and actually become the main effect. Nevertheless, the narcotic effects of such compounds can often be demonstrated if they are used in conjunction with a more purely acting narcotic such as ethyl ether and chloroform. The narcotizing effect of these compounds is then added to the narcotic effects of, for example, ethyl ether, while at the lower concentration that is then sufficient to produce complete narcosis, the above side effects are less perceptible.

This author reached the same theory of narcosis partly through reasoning similar to Meyer's, but in particular through observations that were made in the course of his investigation of the osmotic properties of living plant and animal cells.

During these studies I found that frequently the best, and in some cases the only, way to investigate the relative permeabilities of living cells is detailed study of both the concentration of various narcotics and the time required for narcosis to occur in plants and particularly in animals. A few words will suffice to explain this.

It can very readily be proven by standard experimental methods, such as lack of dehydrating action, that ethyl alcohol, the propyl alcohols, and many other weaker narcotics readily permeate all living plant and animal cells.

For example, if young (8–14 mm long) tadpoles are first placed in a 2% by weight solution of ethyl alcohol, complete narcosis occurs within one or at the most two minutes. If the tadpoles are transferred within 30 minutes or even after 1 hour to pure water, the narcosis becomes incomplete within 2 minutes or less. After 3–5 minutes, the tadpoles are just as lively as before the experiment. In contrast, in 1% by weight solutions of ethyl alcohol, tadpoles never become fully narcotized even after several days, and if tadpoles are transferred from a 2% solution of alcohol into a 1% solution, narcosis fades within a few minutes. Tadpoles in 1.5% solutions become narcotized within 2–3 minutes, and even after being kept for 20 hours in this solution, the narcosis fades again in a few minutes when the tadpoles are returned to pure water.

It can be concluded with almost total certainty from these experiments alone that, if tadpoles are placed in 2% solutions of ethyl alcohol, the concentration of ethyl alcohol in the blood plasma and the water surrounding the ganglia cells of the tadpoles reaches more than 1% in less than one to two minutes. To avoid making any hypothetical assumptions regarding the fine structure of the protoplasm (p. 52), it might be preferable to say that after this short time, more ethyl alcohol from the 2% solution has penetrated the ganglia cells than for any length of time, i.e. at equilibrium, from the 1% solution. It can also be deduced from these experiments that the alcohol leaves the blood plasma and the ganglia cells equally quickly once the tadpoles have been transferred to pure water. It should be kept in mind that, in order to reach the ganglia cells, the alcohol must first penetrate the outer surfaces of the epithelia of the gills into the epithelial cells of the gills, then must leave through their inner surfaces into the space between the epithelia of the gills and the capillaries of the gills. After that, it must pass through the outer surfaces of the capillaries into the body of the endothelial cells and through their inner surfaces into the bloodstream. After being transported by the bloodstream, the alcohol must then pass through the endothelia in the reverse direction in order to reach the intercellular lymph surrounding the ganglia cells, and ultimately penetrate the ganglia cells themselves.

In the case of ethyl alcohol, propyl alcohols, and many other compounds, as indicated earlier, the extraordinarily rapid penetration and loss of alcohol can be detected by methods that are entirely independent of the phenomenon of narcosis. This is not always the case, however, for those compounds that are fatal in even very low concentrations. For example, the instantaneous penetration of isoamyl alcohol into living cells and its immediate loss once the cells are transferred to an isoamyl alcohol free medium can be detected easily both in plant and animal cells by methods that are unrelated to narcosis experiments. This does not apply specifically to penetration of the ganglia cells or a living tadpole as a whole. Since, however, the amount of time for amyl alcohol induced narcosis to both take place and disappear corresponds closely with that for ethyl alcohol, which, in fact, takes place even faster, the very rapid penetration or loss of this compound in the blood and the ganglia can be clearly followed.

Normal octyl alcohol is fatal in such low concentrations that even in the case of plant cells, considerable difficulty exists in detecting its rapid penetration into the undamaged cell by the usual methods, and this can only be achieved with certain particularly favourable test organisms. The demonstration of its absorption by healthy animal cells would be quite impossible using analogous methods, although it can be done readily by observing the course of narcosis. The same condition applies for the numerous other non-specifically acting organic compounds that are toxic at even lower concentrations than octyl alcohol.

Therefore, from the rapid onset of narcosis at concentration c, and its absence at concentration $\frac{3}{4}c$, it can indeed be confidently inferred that the compound penetrates the living cells very rapidly. However, if a compound has a certain particular physiological effect after 10 minutes at concentration c, but has the same effect after longer periods of time at concentrations of only $\frac{1}{2}c$, $\frac{1}{4}c$, or $1/n\ c$, it is not necessarily correct to assume the converse: that such a compound only slowly penetrates the individual cells or the whole body of the organism. This inference would be correct in many cases, but not in others, since equilibrium is not always reached within a short time with respect to the concentration of a foreign compound in the cell, and the physiological response of that cell. In fact, it may sometimes not occur for some time. This must be emphasized even more since some compounds which behave in this fashion in many instances initially produce an effect that is almost indistinguishable from narcosis. This situation applies, for example, to plants and lower animals in the case of hydrocyanic acid and the monovalent aliphatic aldehydes of lower molecular weight. I have characterized compounds of this type that instantaneously penetrate living cells and establish equilibrium between the external and internal cellular concentration, while not achieving their maximum effect until some time later, as progressively acting compounds [2]. I expressed the theory that this phenomenon can be explained by a slow progressive chemical reaction between the corresponding compounds and one or more cellular constituents. Whether the slow effect of these compounds at a given concentration is caused by slow penetration, or the progressive nature of their action (if the concentration of the compound in the cell remains constant) can frequently be determined only by comparative and varied types of experiments.

In the course of these studies of narcosis phenomena, which, as I mentioned, originally served to aid in my research of the relative permeabilities of cells for various compounds, my attention was drawn to a series of regularities that exist between the narcotic strength of a compound and its position in the various groups of chemically related substances, which in turn controls certain physical properties. It turned out, for example, that within the various homologous series, compounds are increasingly stronger narcotics with increasing chain length. This only applies up to a certain length, beyond which the narcotic properties disappear again, or cannot be detected. Furthermore, the strongest narcotic among the various isomers of, for example, an alcohol, is the one whose carbon chain branches the least; i.e. whose molecule is furthest removed from a spherical shape. When comparing the narcotic strengths of benzene, naphthalene, and phenanthrene, it was discovered that phenanthrene had a much stronger narcotic effect than naphthalene and that the latter is much stronger again than benzene. On the other hand, anthracene, which is isomeric with phenanthrene, does not have a noticeable narcotic effect. In addition, if in any given compound, a hydrogen atom or a halogen atom is replaced by a hydroxyl group, the resulting compound was found to have a much lower narcotic strength than the original product. This was even more noticeable when two or more hydroxyl groups were added to the molecule. In contrast, replacing the hydrogen atom of one hydroxyl group by a methyl group, or in more general terms, by an alkyl group, always had the effect of considerably increasing the narcotic strength, or permitting it to be clearly observed. Ths phenomenon could be observed for both an alcoholic and phenolic hydroxyl group. The replacement of a chloride atom by a bromide atom, and a bromide atom by an iodide atom also generally caused an increase in the narcotic strength of the compound.

It seemed to me that all these phenomena, which will be described more precisely in the experimental section (Part Three), together with the particular manner in which the narcotics act, point unambiguously to certain conclusions. First, the narcotic strength of a compound depends primarily upon its partition coefficient between the aqueous and lecithin–cholesterol related dissolving agents within the organism. Secondly, a compound can only be considered an actual narcotic if its absolute solubility in lecithin–cholesterol and similar solvents does not fall below a certain minimum.

All of the above-mentioned regularities with respect to increase in narcotic strength and certain changes in chemical structure are actually related to a shift in relative solubility between water and ether, olive oil, and other solvents. This shift favours the latter group. In fact, this is the only physico-chemical property that changes in a corresponding fashion with respect to increasing narcotic strength, when all of the above modifications are made to the chemical structure.

The fact that in the homologous series, the narcotic strength increases, or appears to increase, with the length of the carbon chain only up to a certain point, can be explained as follows: although the partition coefficients between the solvents water and oil of the consecutive members of the series shift increasingly in favour of the oil, the absolute solubility in oil, lecithin–cholesterol, and other fatty substances at room or blood temperature quickly decreases beyond a certain point within the homologous series. In fact, the decrease is in a more rapid progression that the increase in partition coefficient. To whatever extent the partition coefficient of a compound may favour the olive oil or other fatty solvent with respect to a constantly saturated aqueous solution of the compound, of course, only as much olive oil, ethyl ether, or other fatty substance will be able to pass from the compound into the oil as corresponds to the saturated solution in oil at the relevant temperature. The partition coefficient of cetyl alcohol ($C_{16}H_{31}OH$) between water and oil, for example, undoubtedly favours the oil much more than is the case for octyl alcohol. However, the saturated aqueous solution of cetyl alcohol has no narcotic effect because the absolute solubility of cetyl alcohol in olive oil, lecithin–cholesterol, and other fatty substances at room temperature is too low; the same situation occurs with the higher members of other homologous series. Thus, phenanthrene is readily soluble in cold ethyl ether, olive oil, and other fatty substances at room temperature, while anthracene, which is isomeric with phenanthrene, is not. Therefore, phenanthrene has a narcotic effect, but anthracene does not, even though the partition coefficient between water and ether or water and oil is very favourable to both. I particularly wish to emphasize that even an extraordinarily low solubility of a compound in water does not prevent it from behaving like a narcotic, so long as its solubility even in cold olive oil, or more precisely, lecithin–cholesterol, remains rather high. (Editor's note: subsequent quantitative studies have demonstrated that the decrease in water solubility which is greatly affected by melting temperature can alone account for such a cut off (see R.L. Lipnick (1989) *Environ. Toxicol. Chem.*, **8**, 1).)

Phenanthrene only dissolves in approximately 300 000 to 400 000 parts of water, but produces complete narcosis in solutions of 1 : 1 500 000. However, as soon as the solubility of a compound in water falls below a certain level, for example, 1 : 100 000, narcosis takes place only relatively slowly. In each circulation of the blood plasma, only very slight traces of the compound are transported to the brain. Therefore, the lecithin–cholesterol mixture in the ganglia cells is supplied with the compound only very slowly. A sufficient number of examples of this behaviour will be provided in the specific discussion of the individual narcotics that follows.

REFERENCES

1. Meyer, H. (1899) *Sitzungsberichte der Gesellsch. z. Beförderung der ges. Naturwiss, in Marburg*, Session of January 18; more detail in Meyer, H. Erste Mittheilung: Welche Eigenschaft der Anaesthetica bedingt ihre narcotische Wirkung? *Archiv exp. Pathologie und Pharmakologie*, **42**, 109–18; Baum, F. (1899) Zweite Mittheilung: Ein physikalisch-chemischer Beitrag zur Theorie der Narcotica, pp. 119–24.
2. Overton, E. (1899) Ueber die allgemeinen osmotischen Eigenschaften der Zelle..., *Vierteljahrsschr. Naturforsch. Zürich*, annual edn. **44**, 128 and 129.

Part Two
Experimental Results

4

Narcosis induced by ether and chloroform

Turning now to the discussion of the experimental results with specific narcotics, it seems reasonable first of all to describe in detail the experiments carried out with two compounds and their immediate results. These details should provide the reader with a clear understanding of the progression of experimental events.

I have selected ethyl ether and chloroform for this more thorough presentation partly because of their great practical importance and partly because they represent very typical non-specific narcotics. For most of the remaining typical non-specific narcotics, it will usually suffice to report the test results along with specific notes on the experiments, without describing the experiments themselves in detail.

The individual narcotics are arranged in groups according to their chemical classification. Overall, I have treated individual chemical groups in a similar fashion, as is customary in textbooks of organic chemistry, though I will allow myself some flexibility here.

I considered it expedient to discuss all nitrogen-containing narcotics as long as they are of basic character, separately from the other compounds, although in the mechanism of their action the weakest bases among them scarcely differ from the genuine non-specific narcotics. I did this particularly in order to demonstrate better the gradations from non-specific to basic narcotics.

4.1 ETHER NARCOSIS

Comparative studies demonstrated that there are in fact many gradations between the non-specific narcotics and the various other pharmacological classes of compounds, i.e. antipyretics, excitants and organic antiseptics. I remind the reader also that ethyl ether, chloroform, chloral hydrate and probably all other non-specific narcotics lower the temperature of mammals and birds to some extent and therefore could be designated at the same time as antipyretics. When reviewing the individual organic compounds we will therefore also consider those belonging to the above-mentioned pharmacological groups. Then it will be possible to clarify the difference in the mechanism of action between non-specific narcotics and organic antipyretics.

4.1.1 Experiments with ethyl ether

At 3.41 pm, on May 7, five tadpoles of *Rana temporaria*, 12–13 mm long, were placed in 1%-weight (0.27 mol/l) ethyl ether. After 8–12 seconds, all five tadpoles were already fully narcotized. At 4.11 pm (i.e. after 30 minutes) one of these tadpoles was transferred to fresh water. After 6 minutes it was still not sensitive to stimuli, and even after 8 minutes there was still no discernible circulation in the tail. Twenty minutes later (at 4.40 pm), this tadpole responded to stimuli, but did not yet move spontaneously, and by 4.50 pm it was again quite vigorous and the circulation normal. The four other tadpoles were not transferred to fresh water until 4.21 pm (i.e. after staying for 40 minutes more in the ether solution). By 4.45 pm only one of these four responded to stimuli. By 4.58 pm this one moved spontaneously and the other three responded somewhat to stimuli. By 8.00 pm one had completely recovered, a second still responded to stimuli (it subsequently died, however), and the other two were dead. The temperature during the experiment was 20–21°C.

At 4.27 pm two tadpoles (12–13 mm long) were placed in 1%-weight (0.27 mol/l) ethyl ether. They were transferred after 20 minutes to fresh water and immediately examined under a microscope. Circulation was still present, but extremely slow. They still did not respond to stimuli at 4.50 pm, but both were slightly responsive at 4.55 pm, moving around quite vigorously; the circulation was excellent at 5.00 pm. There were subsequently no signs of harmful after-effects from this extremely strong narcosis.

At 2.30 pm on May 7, four tadpoles of *Rana temporaria* (11–12 mm long) were placed in 0.5%-weight (0.13 mol/l) ethyl ether. All four tadpoles were fully narcotized after only 40 seconds. At 4.15 pm (after 1 hour 45 minutes) two tadpoles in the ether solution were examined under the microscope; no circulation could be detected. At 4.18 pm these two tadpoles were transferred to fresh water. By 4.40 pm one was already moving, but was uncoordinated; the other did not respond. The second one never recovered and the first did not recover completely. At 4.53 pm the remaining two tadpoles were transferred to water, but did not recover and were considered dead before the transfer to water.

At 5.16 pm on April 19, three tadpoles of *Rana temporaria* (10 mm long) were placed in 0.25%-weight (0.07 mol/l) ethyl ether. After 65 seconds all three were fully narcotized. At 7.10 pm all were still completely unresponsive. At 10.00 am on April 20, one tadpole responded somewhat, but the other two did not, although circulation was still maintained. After transfer to fresh water, all three recovered within 1–2 minutes and remained healthy afterwards.

At 5.12 pm on April 19, three tadpoles of *Rana temporaria* were placed in 0.18%-weight (0.05 mol/l) ethyl ether. At 5.14 pm they did not move, but responded somewhat to stimuli, with the same behaviour at 5.19 pm and at 7.10 pm. At 11.45 am on April 20, they moved somewhat when stimulated, and at intervals they moved for a few seconds as if in sleep, without any stimulus. Subsequently, the narcosis decreased somewhat in intensity, although the tadpoles were kept in 400 ml of the solution in a completely air-tight jar that was four-fifths filled with the solution. When transferred after 8 days to fresh water, the tadpoles recovered completely within 1–2 minutes. The circulation was studied under the microscope just before the transfer into water and was found to be good.

These experiments are typical for the course of narcosis at various concentrations for the majority of the non-specific narcotics. In nearly all cases, the narcotics produce death relatively quickly at concentrations significantly higher than those just sufficient to produce narcosis. For tadpoles, death always occurs as a result of cessation of heartbeat. In the case of mammals, the respiratory system seems to be paralysed by most narcotics somewhat sooner than the heart. Although tadpoles are not very useful for studying the effects of narcotics on respiration, the effect on their circulation can be followed easily under the microscope, since weakened heart activity produces complete or partial loss of circulation in the tail. The pulse rate can readily be observed in the tail artery or the arteries of the gills or lips, possibly after removal of the operculum. Young tadpoles of the genus *Bombinator igneus* are so transparent that almost all of the larger arteries and the heart can be observed *intra vitam*. Although all non-specific narcotics have a paralysing effect on the heart at higher concentrations, significant differences exist with respect to the proportions which are just sufficient to produce narcosis and those which produce cessation of the heartbeat within a specified duration. This ratio is of great significance for assessing the utility and risk of the various narcotics, and will thus be considered in the discussion of the individual narcotics.

4.1.2 Calculation of the ether concentration in the blood plasma of narcotized mammals and man from the data of Bert

Earlier, I discussed Bert's experiments on the amount of various narcotics required to produce complete narcosis by inhalation in mammals. At that time I had already indicated that if this amount, which is considered to be constant, is known for a given anaesthetic, along with some physical properties, it is possible to calculate the concentration of the anaesthetic in the blood plasma of the test animal under conditions in which equilibrium has occurred between the level of the anaesthetic in the inspired air and its level in the blood plasma.

The calculation is based upon the application of Henry's Law, or the Henry–Dalton Law, on the absorption of gases and vapours by liquids. According to this law, at a specified temperature, the amount of absorbed gas or vapour is directly proportional to the gas or vapour pressure. Naturally, for gases and vapours absorbed by the receiving liquid in very large amounts, this law applies only up to a certain pressure. In my experience, the law tends to be fairly reliable for vapours until the absorbing liquid has absorbed approximately 6% by weight of the compound. Beyond this level, however, exact proportionality no longer exists between the vapour pressure and the absorbed quantity of compound. Also, at much higher levels, such proportionality is not even approximately correct; we will reconsider this later. We can apply this law to calculate the concentration of ether in the dogs used by Bert as test animals. Let us use specifically those experiments in which the inhaled air mixture contained 20 grams of ether per hectolitre, which is an ether concentration, as noted previously, that is just sufficient to maintain complete narcosis.

Since the molecular weight of ethyl ether is 74, and one litre of hydrogen at 0°C and at normal atmospheric pressure weighs 0.0896 grams, then one litre of ether vapour, assumed to be corrected to 760 mmHg pressure and 0°C, would weigh $37 \times 0.0896 = 3.32$ grams.

The partial vapour pressure of ether at a concentration of 20 grams per hectolitre in the above air mixture, or 0.2 grams per litre, would be $(20/332) \times 760 = 46$ mmHg at 0°C, or $46 \times (293/273) = 50$ mmHg at 20°C, which we will assume to be the test temperature.

Upon inhalation into the alveolae of a warm-blooded animal, the air mixture is warmed to approximately 38°C. However, since the sum of the individual partial pressures of the gases and vapours in the pulmonary vesicles cannot exceed the pressure of the outside air when the inhaled air mixture is warmed, a corresponding expansion takes place, with no net increase in the partial pressure of the warmed ether vapour. (Upon expiration the excess pressure in the alveolae during quiet respiration, such as is present during normal narcosis, hardly exceeds 2–3 mmHg, and even this small amount of additional pressure is compensated for by the slight negative pressure during inspiration.) In fact, as a result of saturating the inhaled air mixture with water vapour, at 38°C the partial pressure of the ether vapour is decreased by about 5%, which we can, however, disregard.

At 38°C, almost exactly 5 grams of ether will dissolve in 100 ml of water. The vapour pressure of pure ether is 840–50 mmHg at 38°C. An aqueous solution that is saturated with ether vapour at this pressure at 38°C, however, generally exists in an unstable form, since the solution tends to separate into two layers, one of which contains a saturated solution of ether in water, and the other, a saturated solution of water in ether. The partial pressure of the ether vapour, as well as the water vapour, is, of course, the same in both phases. It could be readily calculated for the saturated solution of water in ether from the known reduction in ether vapour pressure that results from dissolving 1 mol of any compound in 1 litre. For this purpose information is needed on the molecular weight of water in solutions and the water content of the solution.

Since only about 3% water dissolves in ether and also the water is normally present in solutions in the form of dimers, the reduction amounts to only a few percent of the vapour pressure of pure ether. Thus, we can assume that at 38°C, the partial vapour pressure of ether in a saturated solution of water in ether and therefore also of a saturated solution of ether in water is approximately 810 mmHg.

If 100 ml of water at 38°C and at a partial ether vapour pressure of 810 mmHg dissolves 5 grams of ether, then at the same temperature, and at a partial ether vapour pressure of 50 mmHg, 100 ml of water will absorb $(50 \times 5)/810 = 0.31$ grams of ether.

Therefore, at 38°C, and at a ether partial vapour pressure of 50 mmHg, pure water absorbs slightly less than 0.25% ethyl ether.

A diluted salt solution, and therefore also blood plasma, if fat free, will absorb somewhat less ether than pure water at the same temperature and ether vapour pressure. This, however, represents only a very slight difference in the case of such a diluted salt solution as mammalian blood plasma. With respect to the narcotic effect, the lower concentration of ether in the blood is compensated for exactly by the higher partition coefficient between blood plasma and brain lipoids relative to that of water and brain lipoids. This same relationship, in fact, applies to all non-specific narcotics. All of these are of lower solubility in salt solutions than in pure water, or the partial vapour pressure of these narcotics in a salt solution is always greater at the same temperature than in a solution in pure water at the same narcotic concentration. (In this form, this statement also applies to those narcotics, like the first members of the alcohol series, that are miscible with water in all proportions. It is noteworthy that, while the vapour pressure of a pure solvent is always lowered by the solution of any compound, this need not be the case when a compound is dissolved in a mixed solvent. It is more likely that the sum of the partial pressures of the two components of the solvent will be increased when a compound such as a salt or a carbohydrate is dissolved in it.) Therefore, at the same partial pressure of the narcotics, less of the narcotic will be absorbed from the salt solution than from pure water. In fact, the difference is considerable in the case of concentrated solutions. For example, a 4% solution of sodium chloride absorbs about half as much ether as pure water at the same ether partial vapour pressure. Thus, the partition coefficients of these narcotics is always greater between a salt solution and oil, or lecithin–cholesterol, or other fatty substance, than between pure water and oil. The concentration of narcotics in aqueous solution after they have been distributed is always set as a unit. In addition, the increase in the partition coefficient of narcotics in a saline solution is always related to the decrease in the quantity of narcotic absorbed by the salt solution, so that both changes balance each other exactly. This greatly simplifies experimental work under circumstances in which the narcotic is absorbed by the lung in vapour form. Later, I will briefly discuss the special case in which a non-specific narcotic is absorbed through the gills from a circumambient salt solution or sea water.

4.1.3 Concentration of ether in the blood plasma of narcotized tadpoles

From the above discussion, we may conclude that the concentration of ethyl ether in the blood plasma of a dog that is just sufficient to keep the dog completely narcotized is almost identical to the corresponding blood plasma concentration in the tadpole required to produce narcosis. As Bert and others discovered, the ether level in inspired air needed to narcotize humans is 20 grams of ether per hectolitre. Since the blood temperature of humans and dogs is essentially the same, this leads to the unexpected conclusion that the human brain is just as sensitive to ether as the brain of a dog or tadpole, or at least that the same concentration of ether in the blood plasma is required in all three for complete narcosis of the brain. We will find similar relationships for chloroform. Moreover, I found that with experiments with mice, a mixture containing at least 20 grams of ether per hectolitre was needed in order to produce complete narcosis. Sparrows were found to require an air mixture containing somewhat more ether, corresponding to the fact that the body temperature of birds is several degrees higher than that of mammals. As a result, at the same partial pressure of ether vapour, the blood absorbs somewhat less ether, as discussed below. Experiments with small, warm-blooded animals such as mice and sparrows can be performed without sophisticated apparatus in a large carboy with the basket removed. The experiments can be conducted in the closed carboy containing the air mixture since the composition of this test mixture is not substantially changed when ether is absorbed by these small test animals.

The fact that amphibians require only the same concentration of ether in their blood plasma for narcosis as humans and other mammals is so surprising and conforms so little to earlier theories that it seemed desirable to test its correctness in another way. I did this by the following method.

Two or three small tadpoles 8–10 mm long were placed in a small dish containing 2 ml of water, and the dish was fastened to the end of a long knitting needle with the aid of a strip of wire netting and some wire and placed into a 3 litre wide-mouthed bottle. The other end of the knitting needle was pushed through the tinfoil covered cork of the bottle. With the aid of a capillary pipette, enough ether was introduced into the bottom of the bottle to produce, when evenly distributed, 0.1 grams per litre of ether vapours. The little dish containing the tadpoles in water was moved rapidly up and down by means of the knitting needle (held with tweezers) through the cork. By this means, the dish served as a stirrer which could distribute the ether vapours evenly throughout the entire bottle, while at the same time, the water in the dish became saturated with ether at the corresponding partial pressure of the ether vapours.

As the temperature of this experiment was only 17°C, the tadpoles in the dish became fully narcotized after two minutes, and the narcosis was so intense that the tadpoles expired in less than three hours. At this temperature of 17°C, an ether level in the bottle of only 0.07 grams per litre was sufficient to produce complete narcosis in the tadpoles, whereas, as noted previously, an ether level of 0.2 grams per litre in an inhaled air mixture is needed to maintain narcosis in mammals. Narcosis occurred in tadpoles in our experiment at a much lower partial pressure of the ether vapours than is needed for humans. This is due to the fact that all vapours, like all gases, are absorbed by any absorbing liquid at one and the same partial pressure in greater amounts at a lower temperature than at a higher temperature. Mammals have a blood temperature of about 38°C, while the water in the dish and the blood of the tadpoles were essentially the same in these experiments as the surrounding temperature of 17°C. In fact, calculation shows that a given amount of water absorbs almost exactly the same amount of ether at 17°C from the air mixture containing only 0.07 grams of ether vapours per litre as it does at 38°C from an air mixture containing 0.2 grams of ether per litre. This can be seen from the following data.

At 17°C, approximately 6.7 grams of ether dissolves in 100 grams of water, as opposed to about 5 grams at 38°C. This is because, in contrast to chloroform, ether is less soluble with respect to its vapour pressure in water at higher temperatures than at lower temperatures. On the other hand, pure ether at a temperature of 17°C has a vapour pressure of only about 380 mmHg and the partial pressure of ether in a saturated aqueous solution is somewhat lower at this temperature (17°C), about 360 mmHg. An ether level of 0.07 grams per litre at 0°C corresponds to an ether partial pressure of 16 mmHg, and at 17°C, to a partial pressure of 17 mmHg. Therefore, 100 grams of water at 17°C absorbs from an air mixture whose ether content is maintained at a constant 0.07 grams of ether per litre, $(17/360) \times 6.7$ grams $= 0.32$ grams of ether. By comparison, we found that at 38°C, and with an ether level in the air of 0.2 grams per litre, 0.31 grams of ether were absorbed. The two values are almost identical.

If the same experiment is carried out with tadpoles at 30°C, then the air mixture must contain approximately 0.14 grams of ether per litre in order to produce complete narcosis in the tadpoles. Tadpoles suffer at temperatures exceeding 33–35°C, unless they are allowed to become gradually acclimatized. For this reason, no experiments were conducted at temperatures above 30°C.

4.1.4 Concentration of ether in the blood plasma of other narcotized organisms and in narcotized plants

Even in the case of entomostraca, about the same concentration of ether is sufficient for narcosis as for mammals. This has also been found to be the case so far for other non-specific narcotics. On the other hand, insects – bees, wasps, ants, and flies have been studied – tolerate complete narcosis by ether or chloroform only very poorly, and as a rule, do not recover, at least if the narcosis lasts for some time, i.e. 1–2 hours. Nevertheless, approximately the same concentration of ether in the blood plasma is required for insects as for mammals. However, since in general, the body temperature of insects is only one degree or a few degrees higher than the surrounding temperature – only after a long flight is the difference greater – this means that insects at normal room temperature are narcotized by a much lower ether content in the air than mammals. This can be explained by exactly the same principles that were just set forth for tadpoles. The only difference in this case is that for insects, the ether is absorbed directly from the trachea by the blood plasma of the body lymph, or by the cell tissues, instead of first from the water surrounding the epithelia of the gills and skin, as in tadpoles.

Given this equality in the concentration of ether in the blood plasma required to produce narcosis in mammals, birds, amphibians, insects, and entomostraca, it must again be very strongly emphasized that this is not true for all classes of animals. For the various groups of worms at least twice the concentration of ether is needed for narcosis. For some, even higher concentrations are required. For protozoa and plants, the concentrations of ether required for narcosis are about six times higher than for tadpoles.*

4.1.5 Biological transport of ether and other non-specific narcotics into the blood and cerebral lipoids of aquatic organisms

Finally, before leaving the subject of ether, I must return to the fact that ether (and the same applies to other non-specific narcotics) is less soluble in saline solutions (markedly less in sugar solutions) than in pure water. Furthermore, at a given partial pressure, less ether is absorbed by a saline solution than by the same volume of pure water. As I have already indicated, in animals that breathe through their lungs, the resulting lower concentrations of the narcotic in the blood plasma, which is in equilibrium with a certain partial pressure of the narcotic in the inspired air, is balanced out exactly by the fact that the partition coefficient of the narcotic between the blood plasma and the brain lipoids favours the brain lipoids to a corresponding extent. Therefore, for animals that breathe through gills, attention should be directed to the following, which applies with some modifications in the case of narcotics that are highly soluble in water, to all non-specific narcotics.

If an animal *A*, for example a freshwater fish, is placed in a solution of ethyl ether in pure water with concentration *c*, and a second animal, *B*, for

* In an earlier study (1895, *Vierteljahrsschr. Naturforsch. Ges. Zürich*, annual edn. **40**, p. 34), when I had studied narcosis only in animals that breathe through gills, I thought I was justified in stating that the degree of development in the ganglia cells stood in reverse proportion to the narcotic concentration of various non-specific narcotics that is required to narcotize them. According to the above, this appears to be true only to a certain extent. Perhaps beyond a certain level of development in the ganglia cells, their lecithin–cholesterol content remains the same. I was already aware at that time that entomostraca and tadpoles become narcotized at about the same concentrations, but I saw no reason to assume that the organization of the ganglia cells of tadpoles is higher than that of entomostraca. The degree of development of the nervous system depends on the one hand on the degree of differentiation in the individual ganglia cells, and on the other hand, on the particular manner in which they are related to higher organisms.

example a salt water fish, is placed in a solution of ether of the same concentration, *c*, in a salt solution, for example, sea water or 3.5% NaCl, and if the lecithin–cholesterol content of the ganglia cells of both animals is the same, then, after equilibrium has occurred, the ether level in the ganglia cell lipoids of animal *B* must be greater than the level in the ganglia cell lipoids of animal *A*. The ratio of this difference will be S/S', if S represents the ether concentration in the saturated solution in pure water, and S', the ether concentration in the ether saturated salt water solution. This is true in the case in which the salt level in the blood plasma of the test organisms is essentially the same as that of the surrounding solution. This corresponds to the normal condition of most invertebrate sea organisms. It also applies to those cases in which the saline concentration in the blood plasma of the test organisms is quite different from that of the solution in which the organisms are immersed. The latter situation corresponds to the actual conditions for most marine fish whose blood plasma contains a much less concentrated saline solution than the sea water. For these salt water fish, the osmotic pressure of the blood plasma may be the same as that of the sea water, compensated with respect to osmosis by a urea content, for the lower salt level in the blood, which is the case for cartilaginous fish. Otherwise, the osmotic pressure of the blood plasma may be much lower than that of the sea water, as is the case with most of the *Teleostei* [1]. However, since even a 4% urea content in the water hardly affects the solubility of ether and other non-specific narcotics, it is not a question here of the total osmotic pressure of the blood plasma, but only of its salt content.*

In both cases, the salt content of the blood plasma is lower than in the sea water, and it can be easily demonstrated that, in order to produce equilibrium, enough ether must be transported from the sea water through the epithelia of the gills and the capillary epithelia into the blood plasma so that the concentration of ether in the blood plasma becomes higher than in the sea water surrounding the gills. In fact, the blood plasma concentration must increase to the degree that the solubility of the ether in the blood plasma (the more dilute salt solution) is greater than its solubility in sea water. We have here, therefore, the remarkable case of a compound passing from a region of low partial concentration to one of higher partial concentration. The fact that this actually takes place can be proven by means of Henry's Law, which is just as valid for the absorption of gases and vapours by a salt solution as for their absorption by pure water. This has been shown by Setzschenow for carbon dioxide so long as

* It should be noted that the osmotic solutions of different salts at the same partial pressure of a compound and at the same temperature absorb unequal amounts of the compound, because the individual ions reduce absorption to varying extents.

there is no interfering chemical reaction between the carbon dioxide and the dissolved salt, and by this author for vapours of ether, chloroform, and other compounds. Solutions of ether in water and in a salt solution, or in a more diluted and a more concentrated salt solution, therefore possess only the same ether partial vapour pressure when the ether concentrations of the two solutions are in the same proportion to one another as the concentrations of the saturated solutions of ether in the two solutions. Let us now imagine two cylinders within a hermetically sealed bell jar that are filled with solutions of ether in salt solutions of different concentrations. Furthermore, let us suppose that these cylinders are sealed with semipermeable membranes that are permeable to ether molecules, but not to the molecules and ions of salt or water. It immediately becomes clear that equilibrium can occur only when the partial ether vapour pressure is identical in both cylinders. The gill epithelia of the marine *Teleostei* serve as semipermeable membranes with this same property. If enough urea is dissolved in the cylinder containing the more dilute salt solution to make the total osmotic pressure of the solution equal to the osmotic pressure of the more concentrated solution in the second cylinder, the membrane can be permeable to the water molecules as well as to the ether molecules. In this case, also, a higher concentration of ether in the cylinder containing the more dilute salt solution corresponds to the condition of equilibrium.

These somewhat lengthy explanations were needed to follow what actually takes place when organisms that breathe through gills and have different blood plasma compositions, are placed in a solution of a non-specific narcotic in salt water. The results of our observations can be summarized as follows:

In all these organisms, once equilibrium has been achieved, the lecithin–cholesterol mixture in the ganglia cells can absorb as much ether or other non-specific narcotic as would correspond to its partition coefficient if this mixture were placed directly in the outer salt solution containing the ether. The partition coefficient is always greater than for solutions of ether in pure water. If the salt content of the blood plasma, both qualitatively and quantitatively, is essentially the same as the salt content of the outer liquid, then after equilibrium is achieved, the concentration of ether in the blood plasma of the organism will also be the same as in the outer liquid, and the partition coefficient of the ether between the blood plasma and the lecithin–cholesterol mixture in the ganglia cells becomes the same as that of the ether between the outer salt solution and the lecithin–cholesterol mixture. However, if the salt content of the blood plasma is lower than that of the external aqueous solution, the ether concentration will increase in the blood relative to the external solution, while the partition coefficient of the ether between the blood plasma and the brain lipoids will decrease accordingly, with the net result remaining unchanged.

In fact, it can easily be seen that a certain quantity of oil (or swelled lecithin–cholesterol) must absorb exactly the same amount of ether at a given temperature and ether vapour concentration, regardless of whether the oil, i.e. lecithin–cholesterol, is in direct contact with the ether vapours, or whether they are transported by a system of vessels filled with different solutions, but are all equipped on their contact surfaces with walls permeable to ether molecules. The ether is first absorbed by the vessel that is in direct contact with its vapours, then partially released to more distant solutions, and ultimately to the oil, until the partial ether vapour pressure of all of these solutions is the same, and is in equilibrium with the original ether atmosphere. If this were not the case, this arrangement could produce efficient perpetual motion.

These observations are not only of theoretical interest, but must be taken into account when doing comparative studies. This becomes immediately clear when learning that a 4% salt solution at any partial ether vapour pressure absorbs only two-thirds as much ether as that of pure water at the same partial pressure. In other words, the partition coefficient of ether between a 4% sodium chloride solution and oil (or lecithin–cholesterol) is 50% larger than that between pure water and oil. The difference in the partition coefficients of other non-specific narcotics is regulated similarly. In this regard, sea water is hardly different from that of a 4% sodium chloride solution. There are, of course, some organisms that exist in nature in much more concentrated sodium chloride solutions, and even, according to Darwin [2], in saturated salt solutions. In these cases, the partition coefficients will favour the brain lipoids much more than in the case of a 4% salt solution, and reach several times the value of the partition coefficient between pure water and the brain lipoids.

4.1.6 Partition coefficient of ether between water and olive oil

The numerical value of the partition coefficient of ethyl ether between water and olive oil will now be considered. Only about 6.6 parts by weight of ether dissolve in 100 parts of pure water at 20°C, while ether is miscible in all proportions with olive oil at the same temperature. However, the partition coefficient between water and oil for more dilute solutions, such as used for narcosis, is only about 4.5. This is because Henry's Law applies to aqueous solutions until they are saturated with ether, but not for concentrated solutions of ether in olive oil. Thus, for solutions of about 8%, as the partial pressure of the ether vapour increases, the quantity of ether absorbed by the oil increases more rapidly than the pressure. In a 10% solution of ether in olive oil, the partial pressure of the ether vapours at 20°C is already more than one-third of the pressure of pure ether at the same temperature.

4.2 CHLOROFORM NARCOSIS

4.2.1 Experiments with chloroform

Because of the great importance of chloroform, it seems appropriate, as with ether, to describe in more detail experiments in which it was employed.

At 4.06 pm on April 19 three 9 mm long tadpoles of *Rana temporaria* were placed in 2 litres of a solution of one part by weight of chloroform in 10 000 parts of water (8.3×10^{-4} mol/l). At 4.08 pm they were already very sluggish, but still moved somewhat. At 4.13 pm there were still a few sluggish crawling movements. At 5.10 pm, they still moved somewhat when stimulated, and the same at 7.15 pm. However, when left undisturbed, they remained for hours on the bottom of the container without moving – they did move around very slightly as a result of the movement of the cilia – but they could, however, without stimulus and after rest intervals of several hours, sometimes make movements lasting 2–5 seconds as if in their sleep. Two of the tadpoles remained alive for 10 days in this state, and recovered immediately when transferred to chloroform-free water.

At 4.09 pm on April 19 three 9–10 mm long tadpoles of *Rana temporaria* were placed in 1 litre of a solution of one part by weight of chloroform to 7000 parts water (1.2×10^{-3} mol/l). At 4.13 pm, there were no more spontaneous movements, and at the most, a jerk when stimulated. At 4.20 pm, they did not react to stimuli, but now and again 'spontaneous' jerky tail movements occurred. At 7.20 pm on April 19 they did not respond at all to stimuli. On April 23 the circulation was still good; narcosis was not quite complete. On April 25, they died.

In solutions of 1 : 6000 (1.4×10^{-3} mol/l) complete narcosis occurred and in solutions of 1 : 4000, cessation of heartbeat occurred in a few hours.

4.2.2 Calculations of the chloroform concentration in the blood plasma of mammals from the data of Bert

Now we will calculate the concentration of chloroform required in the blood plasma of mammals and humans for complete narcosis according to Bert's experiments, and using the same principles that have already been discussed for narcosis induced by ether.

In contrast to ether, chloroform is somewhat more soluble in warm water than in cold. At 38°C, 100 grams of water dissolve about 0.78 g of chloroform, as opposed to 0.72 g at 20°C. Since chloroform can dissolve only an extremely small amount of water, the partial pressure of the chloroform vapours in a saturated solution of chloroform in water is practically the same as the vapour pressure of pure chloroform at the same temperature. At 38°C, this vapour pressure is about 338 mmHg. The partial pressure of the chloroform vapours in a solution of 7.8 grams of chloroform in one litre of water (without error, we can also say in one litre of the aqueous solution) is also 338 mmHg at 38°C.

Now Bert found, as already mentioned (p. 41), that in order to narcotize a human or dog completely, the inspired air must contain 8 grams of chloroform per hectolitre. This corresponds to a partial pressure of the chloroform vapours at 0°C of 11.3 mmHg, or 12.1 mmHg at 20°C.

By applying Henry's Law then, it can be determined that at 38°C and a partial chloroform vapour pressure of 12 mmHg, one litre of water absorbs $(12/338) \times 7.8 = 0.275$ grams of chloroform.* Blood plasma, as a saline solution, at the same temperature and partial vapour pressure of chloroform, if completely free of suspended drops of fat, cholesterol, and other lipoids, will absorb somewhat less chloroform. However, this is exactly balanced out by the corresponding greater partition coefficient of chloroform between a salt solution and the brain lipoids. Blood as a whole will absorb more chloroform because of the lecithin–cholesterol content of the blood corpuscles. It is immediately obvious that it is a question here only of the concentration of the chloroform in the blood plasma, and not of the average concentration of chloroform in the blood itself. If drops of fat are suspended in the blood plasma, it naturally also depends only on the concentration of the aqueous solution of chloroform in the blood plasma, while the concentration in the suspended drops of fat and other lipoids does not concern us.

4.2.3 Chloroform concentration in the blood plasma of narcotized tadpoles

It would appear from these figures that a somewhat higher concentration of chloroform is required in the blood plasma of humans and mammals in order to produce narcosis, compared with tadpoles. However, no significance should be attached to this, because, in mammals a fairly long time is required for chloroform to reach complete equilibrium with

* Original formula incorrectly given as $338/12 \times 7.8 = 0.275$.

respect to its level in the inhaled air, the blood, and the tissues. Moreover, it is not impossible that some individual reflexes may still be retained in tadpoles that are maintained in a 1 : 6000 solution of chloroform. With regard to the first point, Dubois, for example, in fact reported that once narcosis has occurred in humans, a chloroform level in the air of only 6 grams per hectolitre would suffice to maintain the narcosis. This would correspond, however, almost exactly to a level in the blood plasma of 1 : 6000 of chloroform. In any case, the differences in the concentrations in the blood plasma required for narcosis to occur in humans and mammals on the one hand, and tadpoles on the other hand, are close to the limit of observation and test error occurring in these studies. If a difference really exists, it lies on the opposite side from that expected according to current theories.

If tadpoles are exposed in a small dish to the vapours of chloroform in the same fashion as described in detail for ethyl ether, it is found that at a temperature of 20°C, a chloroform vapour level of 0.04 grams per litre is already more than adequate to produce complete narcosis. Taking into account the absorption capacity of water, or of blood plasma at this temperature, this again provides proof that the concentration of chloroform in the blood plasma required for narcosis is not substantially different in tadpoles than for humans and mammals. The vapour pressure of pure chloroform at 20°C is 160 mmHg; the partial pressure of chloroform vapours in a saturated solution of chloroform in water at the same temperature is only marginally lower than 160 mmHg. At 20°C, 100 grams of water dissolves 0.0072 g of chloroform. The amount of chloroform that is absorbed from water, or from the blood plasma from an atmosphere containing 0.04 grams per litre can be calculated from these data as before with the aid of Henry's Law. It comes to 0.27 per thousand, i.e. about 1 part by weight of chloroform to 4000 parts of water.

Entomostraca become narcotized at about the same chloroform concentration as tadpoles. On the other hand, the various groups of worms require two to three times this concentration, and protozoa require about six times this concentration.

The partition coefficient of chloroform between olive oil and water is about 30–33 for more dilute solutions, as was found by measuring the partial pressure of chloroform in solutions of oil and water.

REFERENCES

1. Bottazzi (1897) *Arch. Ital. de Biol.*, Part 28, p. 61 and 77; also, this author's notes on this paper in *Vierteljahrsschr. Naturforsch. Ges. Zürich*, 1899, Vol. 44, p. 119, footnote.
2. Darwin, C. *Voyage of the Beagle*, chapter 4.

5

Aliphatic non-electrolyte organic compounds and narcosis

This chapter provides a short overview of the test results for the individual narcotics, partly in the form of tables, and partly in the form of summarizing notes on the different groups of narcotics and related compounds.

5.1 MONOHYDRIC ALCOHOLS

Little needs to be added to Table 5.1. If concentrations of these narcotics are selected that are only about 25% higher than those given in the table, and if tadpoles used for the experiments are only 8–15 mm long, narcosis occurs within about 2 minutes. Only in the case of capryl alcohol does it take somewhat longer, but even with this alcohol narcosis occurs under the conditions described within about 6–8 minutes and, if a concentration of 1 : 8000 is used, in 1–2 minutes. If a concentration is chosen that is very close to the narcotic threshold concentration, narcosis can endure for 24, 36, or more hours, without causing death. After transfer of the organisms to pure water, the narcosis generally disappears in 2–5 minutes, or even sooner. If, however, somewhat more concentrated solutions of narcotics are used, or if narcosis has lasted for a long time, then 30 minutes, an hour, or even longer may be required for narcosis to fade completely after transfer to water, since the circulation has become very weak during narcosis, and as a result, the narcotic is transported very slowly to the epithelia of the gills, and released by them into the surrounding water.

It can be seen from Table 5.1 that the narcotic strength of alcohols of the same structure increases rapidly with molecular weight, and that at the same time, the partition coefficient moves further in favour of the olive oil. Nevertheless, tadpoles maintained in saturated solutions of cetyl alcohol ($C_{16}H_{33}OH$) prepared by shaking the melted alcohol vigorously with warm water, do not become narcotized, even after 5–8 days, even though they become somewhat sluggish after a few hours. This sluggishness decreases rather than increases after the first day. The undissolved cetyl alcohol remains partially in a very finely divided state of

Table 5.1 Monohydric alcohols*

Compound	*Concentration for complete brain narcosis in tadpoles*		*Partition coefficient (olive oil/water) or solubility in both solvents*
	(parts by weight of water per one part by weight of narcotic)	*(mol/l)*	
Methyl alcohol	50–60	0.52–0.62	Miscible with water; <1 : 50 in oil
Ethyl alcohol	70–80	0.27–0.31	Partition coefficient about 1/30
n-Propyl alcohol	150	0.11	Partition coefficient about 1/8
iso-Propyl alcohol	130	0.13	—
n-Butyl alcohol	350	0.038	Soluble in 12 parts of water; miscible with oil
iso-Butyl alcohol (of fermentation)	300	0.045	Partition coefficient about 6; soluble in 10 parts of water
tert-Butyl alcohol (trimethyl carbinol)	100	0.13	Miscible with water and oil; partition coefficient much less in favour of oil than the last two
Amyl alcohol (of fermentation) [Editor's note: iso-amyl alcohol]	500	0.023	2% solubility in water; miscible with water

Table 5.1 *Continued*

Compound	*Concentration for complete brain narcosis in tadpoles*		*Partition coefficient (olive oil/water) or solubility in both solvents*
	(parts by weight of water per one part by weight of narcotic)	*(mol/l)*	
Amyl hydrate (dimethyl ethyl carbinol) [Editor's note: tert-amyl alcohol]	200	0.057	Soluble in 8 parts of water; miscible with oil
Capryl alcohol ($C_8H_{17}OH$)	20 000	0.0004	Soluble in about 2000 parts water; miscible with oil
Cetyl alcohol ($C_{16}H_{38}OH$)	No narcosis (saturated solution)		Extremely little solubility in water; rather poorly soluble in cold oil
Ethyl mercaptan† (C_2H_5SH)	2000	0.008	About 1% solubility in water; miscible with oil

* Editor's note: see reference 11, Appendix C, for more recent measurements of the olive oil/water partition coefficient for these and other compounds and for octanol/water partition coefficient values.

† Not a saturated monohydric alcohol but included by Overton as part of this table.

suspension, so that the solution always remains saturated. Undoubtedly, the partition coefficient between water and oil also increases greatly from capryl (octyl) alcohol to cetyl alcohol. However, the absolute solubility of cetyl alcohol in cold oil and in the cold lecithin–cerebrine mixture of the ganglia cells is too low for narcosis to be induced by the saturated solution. A solution of ether in water that is not quite sufficiently concentrated to produce complete narcosis alone, causes narcosis if cetyl alcohol is present in the solution to saturation. The very low solubility of cetyl alcohol in water, which can be in the order of 1:500000, is of significance only in so far as the onset of practical equilibrium between the tissues and the outer solution is delayed. However, everything points to the fact that this equilibrium is essentially reached within one day. The even higher alcohols, e.g. ceryl alcohol ($C_{27}H_{55}OH$) and melissyl alcohol ($C_{30}H_{61}OH$), have just as little narcotic effect as cetyl alcohol, and for the same reasons. The alcohols $C_9H_{19}OH$ to $C_{15}H_{31}OH$ are unfortunately not widely available. The saturated solution of the alcohol $C_{14}H_{29}OH$ probably still produces narcosis.

In the case of the isomeric alcohols, the normal ones produce narcosis at lower concentrations than the isomeric alcohols with a branched chain. In fact, the more branched the chain, the greater must be the concentration, with tertiary alcohols always having the lowest narcotic strength. In this case, as well, this is related to the partition coefficient, since among the various isomers, the normal alcohols are the least soluble in water, and the tertiary alcohols are the most soluble in water, and the partition coefficient between water and oil in the first case favours oil the most, but in the second case, favours oil the least.

It is indeed noteworthy that methyl and ethyl alcohol produce narcosis in plant cells, protozoa, ciliary cells, and others at concentrations that are only two to three times higher than the concentration required to produce narcosis in tadpoles, while the higher alcohols narcotize plant cells only at 8–10 times that concentration. This indicates that the action of ethyl and methyl alcohol, at least on cells of lower order, is a mixed one, and that the action extends to other constituents of the protoplasm besides the lecithin and cholesterol-related substances in the cells. Similar relationships can be found in the lowest members of the other homologous series, e.g. ketones, if these are more soluble in water than in oils.

5.2 ALIPHATIC HYDROCARBONS AND THEIR HALOGEN DERIVATIVES

Equilibrium is reached with respect to the concentration in tissues and in the surrounding solution just as rapidly for these compounds as for those belonging to the first group. Narcosis also fades just as quickly when the tadpoles are transferred to fresh water. The exact threshold concentration is often difficult to determine, since the tadpoles appear to be unresponsive for hours in a solution maintained at a constant concentration, but sometimes move spontaneously and then become responsive again. Nevertheless, this behaviour, which is manifested mainly for compounds of the type C_2H_5X (Table 5.2), is observed only within a rather narrow range of concentration.

Table 5.2 Aliphatic hydrocarbons and their halogen derivatives

Compound	*Concentration for complete brain narcosis in tadpoles*		*Solubility in oil and water*
	(parts by weight of water per one part by weight of narcotic)	*(mol/l)*	
Pentane	5000	0.0028	About 1 : 2000 solubility in water; miscible with oil
Amylene	6000	0.0023	About 1 : 1700 solubility in water; miscible with oil
Ethyl chloride	3000–4000	0.004–0.005	Solubility in water almost exactly 1%; miscible with oil
Ethyl bromide	3000–4000	0.0023–0.0031	Solubility in water almost exactly 1%; miscible with oil
Ethyl iodide	6000	0.0011	Solubility in water 0.25%; miscible with oil
Ethylene chloride*	4000–4500	0.0022–0.0025	—
Chloroform	6000	0.0014	0.72% by weight solubility in water at 20°C; miscible with oil. Partition coefficient 30–33
Tetrachloromethane	8000–10 000	0.00064–0.0008	—

* Editor's note: ethylene dichloride.

Aqueous solutions of the lowest hydrocarbons, such as methane and ethane, that are saturated at normal atmospheric pressure, have no apparent narcotic effect. According to estimates based upon observations of the regularities in the increase in narcotic strength when certain substitutions are made, it can be predicted with fair probability, for example, that methane will produce complete narcosis in tadpoles at 10°C only at a pressure of 18–30 atmospheres.

Narcosis induced by chlorine derivatives can be maintained much longer, without causing death, than when bromine and iodine derivatives are used. This difference probably results from the fact that aqueous solutions of the bromo- and iodo- hydrocarbon derivatives gradually split off bromine and iodine. I was not able to induce narcosis by means of iodoform; in solutions of approximately 1 : 20 000, tadpoles died within 12–20 hours without the occurrence of narcosis.

Entomostraca become narcotized at the same concentration as tadpoles; worms require two to three times the concentration; and protozoa and plant cells need six times the concentration.

Binz [1] considers that the mechanism of narcotic action of halogenated hydrocarbons is clearly unrelated to the action of bromides. Tadpoles can be maintained for a week or more in a 1% solution of sodium bromide (NaBr) without showing any apparent effects, and the solutions do not appear to penetrate. In solutions of 1.5% or more of NaBr, tadpoles become gradually dehydrated, as occurs with all non-penetrating solutions that possess an osmotic pressure of more than 4 atmospheres. The same applies to sodium iodide (NaI) for which solutions have no narcotic effect. The potassium compounds KCl, KBr, and KI have a gradual toxic effect. In this case, however, it is a question only of the specific effect of the potassium. The effect does not become noticeable for hours or even days, since these salts are absorbed only slowly and probably only by the intestinal canal. Solutions of free iodine of 1 : 1 000 000 cause death in 15 mm long tadpoles at about 20°C within about 4 hours. During the first hour, no effects are apparent, but after 2 hours, both movement and the epithelia are rather well affected, with circulation usually well maintained, but weakened later. In any case, free iodine has a fairly uniform effect on all tissues. As opposed to Binz, I would not ascribe any narcotic effect to it [2].

5.3 NITRILES AND NITROPARAFFINS

Among the nitroparaffins, only nitromethane was tested more carefully; from the nitriles, only acetonitrile. Nitromethane produces complete narcosis in tadpoles at a concentration of 0.5% (0.082 mol/l). Following a narcosis of 5 hours duration, the animals recovered completely. Nitromethane dissolves in approximately 9.5 parts water and in 8.5 parts oil.

Acetonitrile produces an effect very similar to ethyl alcohol, although based upon the calculated molecular concentrations, it is a somewhat weaker narcotic, since complete lack of response does not occur below 1% by weight (0.36 mol/l). Prior to the onset of narcosis, there is a certain tendency for spasms; tadpoles can survive for days in 1% solutions. Acetonitrile is miscible with water and dissolves in approximately 40 parts of olive oil.

Hydrocyanic acid acts quite differently from alkyl nitriles. A narcosis-like state is produced with much more dilute solutions after a lengthy test duration. This is observed despite the fact that hydrocyanic acid penetrates the tissue very rapidly so that an equilibrium with respect to the concentration in the cell tissues and the surrounding solution is reached very quickly. Its mechanism of action is very different from that of normal narcotics, in that the action is progressive, and probably depends upon a slow chemical reaction between the hydrocyanic acid and some constituent of the protoplasm. At any concentration, the narcotic state disappears after transfer of the tadpoles to fresh water only if the condition had persisted for a short time.

5.4 MONOVALENT ALDEHYDES, PARALDEHYDE, CHLORAL HYDRATE, AND CHLORALFORMAMIDE

The first members of the monovalent aldehydes, such as formaldehyde and acetaldehyde cannot be considered true narcotics since, as in the case of hydrocyanic acid, a highly progressive effect is observed. This occurs less often with the higher homologues, but still is evident to a certain extent. The fact that aldehydes undergo oxidation to the corresponding acids in oxygenated aqueous solutions further complicates their behaviour in that the action of an acid is quickly apparent and adds to the effect of the aldehyde. In the case of paraldehyde, which is actually not an aldehyde, chloral hydrate and its derivative chloralformamide, these complications are absent, or recede into the background entirely.

Paraldehyde ($C_2H_4O_3$)
Narcotic threshold solution 1 : 300 (0.025 mol/l). Paraldehyde is soluble in 10 parts of water and is miscible with oil; its partition coefficient is approximately 3.

A solution of paraldehyde that has been stored for several years in the dark quickly causes death even at concentrations that are not sufficient for narcosis. With rather pure paraldehyde, narcosis can last about 20 hours without causing death, but the concentration may not exceed the narcotic threshold concentration. A 0.5% solution causes death within 2–3 hours; in these solutions, narcosis in smaller tadpoles takes place within 1 minute.

Chloral hydrate

This compound penetrates much more slowly through the epithelia of the gills and the capillary epithelia into the blood and from there into the cell tissues than all the narcotics discussed up to now. Partly for this reason, and partly for other reasons, it seems to me expedient to describe the experiments in somewhat greater detail.

At 4.04 pm on April 13 1894, three tadpoles of *Rana temporaria* (16 mm long) were placed in 0.25% chloral hydrate. At 4.25 pm (after 21 minutes) they were moving normally; at 4.40 pm, movement was sluggish; and at 4.58 pm, they were asleep, but still responded to stimuli, and even moved around when stimulated. At 6.00 pm, they were unresponsive; the same at 7.30 pm. At 7.50 pm, they were transferred to fresh water; at 9.00 pm, they were still completely unresponsive, but the following morning were quite vigorous.

Even in 0.5% solutions of chloral hydrate, narcosis does not occur for 15 minutes, with an even longer delay for larger tadpoles. If the tadpoles are left for another 10–15 minutes in the solution prior to being transferred to fresh water, 2–3 hours are required before the narcosis fades, since the chloral hydrate is released rather slowly from the tissues and the blood into the surrounding water. In solutions of 1 part per 1000, narcosis occurs in 4 hours, if at all. The tadpoles of the various types of amphibia do not behave exactly the same with respect to the concentration finally producing narcosis. Nevertheless, this concentration lies somewhere between 1 : 800 and 1 : 1200 (5.1×10^{-3}–7.6×10^{-3} mol/l). Tadpoles live for days in concentrations of 1 : 2000 and at the most, become somewhat sluggish.

Chloral hydrate is much more soluble in water than in oil; the partition coefficient is approximately 0.22 (Baum).

We know that Liebreich attributed the effectiveness of chloral hydrate to cleavage by the alkalis of the blood into chloroform and formic acid. In this case, it was assumed that chloroform alone represented the active substance. Until becoming more involved with the literature on this subject, I considered that this hypothesis had been completely abandoned again. In the meantime, in France, it has become, in somewhat modified form, the prevailing theory, as presented in treatises on narcotics of Dastre (1890), R. Dubois (1894), and others. In the modified form of the hypothesis, which was proposed by Arloing [3], actual narcosis by chloral hydrate is attributed solely to the chloroform cleavage product, whereas its effect on the vascular system, respiration, and elsewhere is attributed to the combined action of chloroform and formate.

To support his hypothesis, that narcosis results from the chloroform produced, Arloing argues that plants (mimosa) are narcotized by chloroform, but not by chloral hydrate, and he explains this by stating that plant fluids are acidic, and therefore do not provide the required conditions to split the chloral hydrate. This argument completely falls apart upon the realization that the assertion that plants are not narcotized by chloral hydrate is entirely incorrect, and can only be explained by Arloing's unlucky choice of test plants and his mistaken experimental method. I discovered many years ago that all plant cells (spirogyra, nitella, chara, elodea, root hairs of hydrocharis, and others) are narcotized by 0.5–0.75% chloral hydrate, and in fact, all protoplasm flow ceases, as does carbon dioxide decomposition in light. There is absolutely no likelihood of a perceptible cleavage of chloroform in this case; besides which, plant cells are only narcotized by 1 part per 1000 of chloroform. There is also the fact that chloroform enters and leaves the living cells much more rapidly than chloral hydrate. Whether in the blood or in the cell tissues of tadpoles, chloroform formed from chloral hydrate would have to pass through the epithelia of the gills and skin into the surrounding medium as soon as it is produced, without the chloroform in the blood plasma and the ganglia cells being able to reach even remotely the concentration required for narcosis. Moreover, also taking into account the slight alkalinity of amphibian blood and its relatively low temperature, the shakiness of the splitting hypothesis becomes entirely obvious. There is no doubt that chloral hydrate is in itself effective.

Once the issue of cleavage is no longer under consideration, it does not seem improbable to me that the mechanism of action of chloral hydrate is more complex than in the case of most other non-specific narcotics. Its action may extend to the protein substances of the cell, as well as to the lecithin–cholesterols. Chloral hydrate is, in fact, thought to form a compound with proteins, although these compounds have been studied very little up to now. I remind the reader also that solutions of chloral hydrate always have a somewhat acid reaction. If chloral hydrate truly forms a compound with cell proteins, it should probably be treated with an analogous situation that we shall encounter later in the case of phenols.

Chloralformamide

Its narcotic threshold concentration is approximately 1 : 300 (0.017 mol/l). Here there is no doubt that chloralformamide itself can have a narcotic effect. Since it can split very easily into chloral and ammonium formate, however, it is not unlikely that in mammals in the later stages of narcosis enough chloral hydrate is split off to play a part in the narcotic action.

5.5 KETONES, SULFONALS, ALDOXIMES AND KETOXIMES

The lowest member of the ketones (Table 5.3), acetone, can hardly be considered a narcotic, since complete narcosis does not take place until a concentration of 1.5% by weight (0.26 mol/l) is used. However, at this concentration, death follows within a short period of time from cessation of the heartbeat. In 0.8% solutions, responses were retained, even after 3 hours, but the circulation was greatly weakened, and death followed within 7 hours. In a 0.4% solution, the tadpoles maintained movement for 14 hours; after 19 hours, however, they were all dead. Acetone is miscible with water and oil, but on the other hand, is difficult to dissolve in olive oil.

Methyl ethyl ketone is a much better narcotic, but it also causes death relatively quickly. All of the other monovalent ketones are very good narcotics. Tadpoles, crustaceans, and other organisms remain alive for days in solutions of these substances, whose concentrations are not quite sufficient to induce narcosis. All ketones penetrate the tissue very rapidly, so that solutions of them in the tissues and in the outer fluid reach equilibrium within essentially a few minutes. Thus, hypnone (methyl phenyl ketone) at a concentration of 1 : 4000 narcotizes in 3 minutes, while a concentration of 1 : 12 000 never leads to complete narcosis.

Complete narcosis of longer duration cannot be achieved in tadpoles with sulfonal and trional, since the circulation is too strongly affected by these compounds. Tadpoles were still responsive to stimulus in one experiment after 6 hours and 30 minutes in 1 : 500 sulfonal.*

However, the pulse had dropped to about 50 beats per minute (instead of 80–100) and 2 hours later, circulation stopped. In a second experiment at the same concentration, circulation ceased after 5 hours and 30 minutes. Tadpoles in 1 : 600 trional still responded just perceptibly to stimuli after 1 hour 45 minutes. After another 3 hours and 30 minutes, no further circulation could be observed.

Acetoxime and to a lesser degree acetaldoxime affect the circulation after a few hours, so that narcosis cannot last long without producing death.

5.6 ESTERS OF MINERAL ACIDS

From among the mineral acid esters, ethyl nitrate, triethyl phosphate, and dimethyl sulphate were studied.

* For further information on the sulfonal group see Baum (1899) Ein physikalisch-chemischer Beitrag zur theorie der Narcotica, *Arch. f. exp. Pathol. und Pharmakol.*, **42**, 119–24; and Diehl (1894) Vergleichende unter suchungen uber die starke der narcotischen Wirkung einiger Sulfone, Saureamide und Glycerinderivate, *Dissertation*, Marburg. In the first study, the partition coefficients are also given between water and oil. (Both studies were carried out under the guidance of H. Meyer.)

Table 5.3 Ketones, sulfonals, aldoximes and ketoximes

Compound	*Concentration for complete brain narcosis in tadpoles*		*Solubility in oil and water*
	(parts by weight of water per one part by weight of narcotic)	*(mol/l)*	
Acetone	60 (see text)	0.26	Miscible with water; slightly soluble in oil
Methyl ethyl ketone	150	0.9	Soluble in 5 parts of water; miscible with oil
Diethyl ketone	400	0.029	Soluble in 20 parts of water; miscible with oil
Methyl propyl ketone	600	0.019	Soluble in 25 parts of water; miscible with oil
Hypnone (methyl phenyl ketone)	8000–10 000	0.00083–0.00104	Soluble in 330 parts of water; miscible with oil
Sulfonal	550 (responds to stimulus but immobile)	0.0088	Soluble in about 500 parts of cold water, in 110 parts of oil
Trional	600 (responds to stimulus but immobile)	0.0069	Soluble in about 320 part of cold water, in 20 parts of oil
Acetaldoxime	130–160	0.13–0.11	—
Acetoxime	160–200	0.086–0.068	—

Ethyl nitrate produces narcosis in 1 : 1500 (0.0073 mol/l), which can last more than 24 hours without producing death. During the entire period of narcosis, the circulation remains extraordinarily strong, so that after the tadpoles are transferred to fresh water, the narcosis disappears extremely quickly. The water solubility of ethyl nitrate is slightly more than 1%, and it is miscible with oil. Solutions of completely pure ethyl nitrate demonstrate a weak acid reaction after several days, but there is scarcely any in the first 2–3 days.

Triethyl phosphate
This compound was manufactured upon my request by the E. Merck factory in Darmstadt, and delivered in a highly pure state. In 1 : 500 (0.011 mol/l), tadpoles stopped moving after 20 minutes, but still responded to stimulus. After 30 minutes, however, they no longer responded. Circulation was still good after 3 hours. Following a narcosis of 18 hours, five tadpoles recovered completely after being transferred to fresh water. Tadpoles remained alive in 1 : 1000 solutions for more than 50 hours, merely becoming rather sluggish, but still moving around somewhat at the end of the experiment. Triethyl phosphate is miscible with water. The oil/water partition coefficient is 1/12 for concentrated solutions, which is an unusually low number for a compound of the 'narcotic potency' of this ester.

Dimethyl sulphate
When dissolved in water, dimethyl sulphate momentarily undergoes fairly strong hydrolysis, so that in the first minute, the solutions turn blue litmus a deep red. Accordingly, this compound has a fatal effect before narcotic symptoms can appear.

5.7 ESTERS OF ORGANIC ACIDS: SIGNIFICANCE OF THE RATE OF SAPONIFICATION, AND EFFECT OF THE PRESENCE OF HYDROXYL GROUPS

This group of narcotics is interesting in several respects and is very important for the theory of narcosis. For the sake of simplicity, we can distinguish three sub-groups: (1) esters of monovalent acids, (2) urethanes, and (3) esters of polybasic acids. The most important data are presented in Table 5.4 which is supplemented with additional information.

It is immediately apparent from Table 5.4 that the narcotic strength of esters of monobasic acids increases with molecular weight, while their relative solubility in water and oil increasingly favours the oil. The esters

Table 5.4 Esters of organic acids

Compound	*Concentration for complete brain narcosis in tadpoles*		*Partition coefficient (olive oil/water) or solubility in both solvents*
	(parts by weight of water per one part by weight of narcotic)	*(mol/l)*	
Methyl acetate	150–200	0.07–0.09	—
Ethyl formate	200	0.07	Oil/water = 4
Ethyl acetate	400	0.03	Soluble in 15.2 parts of water; miscible with oil
Ethyl propionate	800–1000	0.0098–0.012	Soluble in 50 parts of water; miscible with oil
Propyl acetate	800–1000	0.0098–0.012	Soluble in 50 parts of water; miscible with oil
Ethyl *n*-butyrate	2000	0.0043	Soluble in 190 parts of water; miscible with oil
Ethyl *iso*-butyrate	1500	0.0057	Soluble in 140 parts of water; miscible with oil
n-Butyl acetate	1500–2000	0.0043–0.0057	Soluble in 180 parts of water; miscible with oil
iso-Butyl acetate	1500	0.0057	Soluble in 180 parts of water; miscible with oil
Ethyl valerate	4000	0.0019	Soluble in 500 parts of water; miscible with oil
Amyl acetate	4000	0.0019	Soluble in 500 parts of water; miscible with oil

Table 5.4 *Continued*

Compound	*Concentration for complete brain narcosis in tadpoles*		*Partition coefficient (olive oil/water) or solubility in both solvents*
	(parts by weight of water per one part by weight of narcotic)	*(mol/l)*	
Butyl valerate	25 000	0.00025	Soluble in 3500 parts of water; miscible with oil
Methyl urethane	50	0.27	—
Ethyl urethane	250–300	0.037–0.045	Soluble in 1 part of water; soluble in 20 parts of oil
Phenyl urethane	10 000–12 000	0.0006–0.0007	Soluble in 720 parts of water; soluble in 3.5–4 parts of oil
Tartaric acid, ethyl ester	80	0.061	Miscible with water; soluble in about 70 parts of oil
Citric acid, ethyl ester	400	0.009	Soluble in 25 parts of water; miscible with oil

$C_mH_{2m+1} \cdot C_nH_{2n-1}O_2$ and $C_nH_{2n+1} \cdot C_mH_{2m-1}O_2$ possess the same narcotic strength so long as both carbon atoms have the same structure. The relative and absolute solubilities of such pairs in water and oil, are in fact, identical. On the other hand, a comparison of the two esters of the same molecular weight, but of different carbon chain structure, reveals that the ester with the non-branching, or less branched chain always has a noticeably higher narcotic strength. This is related to the fact that the compounds with more highly branched chains and the same number of carbon atoms are more soluble in water and have an (oil/water) partition coefficient that favours oil less.

Furthermore, several facts not shown in Table 5.4 are of great interest regarding the induction of narcosis by these esters. It is of interest to compare the behaviour of esters having a common acid moiety but varying alcohol constituents, with those esters having a common alcohol moiety but varying acid constituents. Such a comparison reveals that the longer the carbon chain of the alcohol or acid moiety, the longer narcosis can be sustained without fatal side effects. However, in this connection the influence of the carbon chain length is of greater importance for the acid moiety than the alcohol moiety. Furthermore, temperature also plays an important role. A few examples below illustrate this point.

At a temperature of about 20°C, narcosis induced by ethyl formate causes death if it lasts longer than 20–30 minutes, whereas narcosis induced by methyl acetate can last about 35–45 minutes. At the same temperature, narcosis by amyl acetate can last only about 2½–3 hours, and narcosis by ethyl valerate can last for some 15 hours, without being fatal. At a temperature of 5°C, death follows much more slowly than at 20°C, and at 20°C it occurs later than at 30°C. All of these phenomena are related to the unequal rate of saponification of the different esters. Death caused by these esters is in almost all cases a result of hydrolysis of the ester to the free acid and the corresponding alcohol, and is not a direct consequence of narcosis. In fact, death occurs in a relatively short time, even when the solution of the particular ester is not sufficiently concentrated to produce narcosis. Investigations by physical chemists [4] have in fact demonstrated that the rate of saponification decreases with the length of the carbon chain of both the acid and alcohol moieties, but that the length of the acid moiety exerts the greatest influence. It is common knowledge that the saponification rate increases with temperature as do the rates of all chemical reactions. It should also be noted that in all of my experiments with the esters of the monobasic acids, at the beginning of the experiment the solution was made just perceptibly alkaline by the addition of a small amount of Na_2CO_3. Saponification appears to take place more quickly in the fluids and tissues of the tadpole test organisms than in the surrounding solution.

A concentration of the ester that is approximately 6–12 times greater than for tadpoles is required to produce complete narcosis in plant cells. The relative difference is greatest for those esters having long carbon chains. Plant cells are much less quickly damaged by the free acid resulting from the saponification than are most organisms.

It should be added to the data on the urethanes that an extraordinarily long lasting narcosis can be maintained in tadpoles by means of phenyl urethane without fatality, even in concentrations that greatly exceed the narcotic threshold, with circulation remaining unaffected for a long time. Thus, tadpoles of *Bufo variabilis* remained for over 80 hours in 1 : 12 000 phenyl urethane and were completely unresponsive to stimuli, but recovered immediately after being transferred to fresh water. The circulation was studied several times during the course of the experiment and remained relatively strong for the entire time. In a 1 : 6500 solution, i.e. one having twice the concentration required for narcosis, a tadpole remained alive for 26 hours, and another for 36 hours, although the pulse rate dropped to 40–45 beats per minute.

In the case of the neutral ethyl ester of citric acid and the neutral ethyl ester of tartaric acid, the first of which contains nine carbon atoms, and the second, eight, it is noteworthy that these possess a much lower narcotic strength than the esters of monobasic acids with the same number of carbon atoms. This is because the addition of a hydroxyl group to a molecule always tends to increase the solubility of the resulting compound in water. In contrast, it also has the effect of lowering its solubility in ether, oil, and other fatty substances, or at the very least, of shifting the partition coefficient of the compound between water and ether, or between water and oil, more in favour of the water. With the addition of two or more hydroxyl groups, this shift in the partition coefficient in favour of water takes place to an even greater degree. The neutral ethyl esters of tartaric acid and citric acid are saponified only very slowly by water. After three or four days, the solutions still have no noticeable acid reaction, and accordingly, at 20°C, narcosis can last for some 20 hours without adverse effects. Tadpoles survive more than 8 days in 0.5% solutions of ethyl tartarate. With this ester, approximately an hour is required before equilibrium is reached in smaller tadpoles (12–15 mm) with respect to the concentration of the ester in the tissues and in the surrounding solution. In the case of the citric acid ester, equilibrium is reached more rapidly.

The esters of oxalic acid are saponified very quickly in water, so that they are fatal within a few minutes. The ethyl ester of succinic acid is saponified more slowly, but its use also produces cessation of heartbeat either simultaneously, or prior to the induction of narcosis.

5.8 DIHYDRIC AND POLYHYDRIC ALCOHOLS AND SOME OF THEIR DERIVATIVES

The di- and trihydric alcohols have a much lower narcotic strength than the monohydric ones. In fact, the lower members of these series do not produce cerebral narcosis. Even at lower concentrations, they are more harmful to certain tissue constituents than to the ganglia cells. This is related to the condition noted above that the addition of hydroxyl groups to a molecule shifts the oil/water partition coefficient strongly in favour of water. The higher members, at least of the dihydric alcohols, do have a narcotic effect, since the influence of the hydroxyl group is weakened in this instance by the opposing effect of a longer carbon chain.

Ethylene glycol, for example, which is soluble in water in all proportions, but only very slightly soluble in ether and olive oil, does not produce narcosis in tadpoles even in 4% solutions (0.46 mol/l). The tadpoles become sluggish after 5–6 hours, but they do not lose their response to stimuli. The sluggishness of their movements appears to be more the result of a direct effect of glycol on the muscles, particularly on the heart musculature. Death follows within 12–20 hours as a result of progressive weakening of the circulation. In 2% solutions, tadpoles keep moving for about 24–30 hours, but die after about 36–48 hours. Ethylene glycol penetrates tissues much more slowly than monohydric alcohols, and can produce a fairly damaging degree of dehydration if the test organisms are transferred directly from fresh water into fairly strong solutions.

Pinacol $(CH_3)_2C(OH)C(OH)(CH_3)_2 \cdot 6H_2O$* produces narcosis in 4–5% solutions (0.14–0.17 mol/l). This compound, despite the presence of six carbon atoms, is still very soluble in water, but dissolves with difficulty in ether and oil.

Glycerine penetrates the epithelia of the gills and skin, entering the blood plasma and the lymph of tadpoles very slowly, so that the cell tissues, which are themselves only slowly permeated by glycerine, require 24–36 hours to absorb a sufficient amount of glycerine to reach equilibrium with the external solution. If tadpoles are placed directly in glycerine solutions of more than 1.75%, a certain amount of dehydration of the blood and tissues takes place. In order to avoid this, it is necessary first to place the tadpoles in 1.5% solutions, and then very slowly increase the concentration by approximately 0.5% every 12–20 hours. I, myself, have never succeeded in producing narcosis even by this method. The tadpoles always died from other causes – apparently cessation of the heart – before the concentration of glycerine needed for brain narcosis was reached in the ganglia cells. Glycerine is only faintly soluble in ether and olive oil. In solutions of slightly greater than 20%, glycerine produces narcosis in plant cells, ciliary cells, and others. Nevertheless, to avoid dehydration when experimenting with these cells, the concentration should gradually be elevated to this level.

However, if one hydroxyl group in ethylene glycol or glycerine is replaced by a halogen atom, or the hydrogen atom of the hydroxyl group is replaced by an alkyl group, the resulting compound again exhibits rather pronounced narcotic properties. The degree of these properties depends upon the nature of the substituting atom or group, and, at the same time, the compound penetrates the blood and tissues more rapidly as the partition coefficient between water and oil shifts further in favour of the oil. If two or, in the case of glycerine, all three hydroxyl groups are

* Editor's note: formula incorrectly given by Overton as $(CH_3)_2 \cdot CH(OH) \cdot (CH_3)_2 \cdot CH_2OH + 6H_2O$.

replaced by halogen atoms, or the hydrogen atoms of the hydroxyl groups are replaced by alkyls, the resulting compounds have a high narcotic potency, penetrate the blood and tissue very rapidly, and have a partition coefficient that favours oil even more.

For example, monochlorohydrin ($CH_2(OH)CH(OH)CH_2Cl$), still has a very low narcotic strength, since it narcotizes tadpoles in a solution of only 1.75–2% by weight (0.16–0.18 mol/l). A narcosis of more than 3 hours duration is usually fatal. On the other hand, in 1% solutions, tadpoles only become a little sluggish, and they can survive for 2–3 days. Monochlorohydrin is miscible with water, but soluble in oil only to about 10%.

Dichlorohydrin ($CH_2ClCH(OH)CH_2Cl$), which is soluble only in 9 parts of water, but miscible with oil, narcotizes tadpoles in 1 : 600–1 : 800 (0.01–0.013 mol/l).

Glycerine monoethyl ether $C_3H_5(OC_2H_5)(OH)_2$, which is readily soluble in water, however, acts as only a weak narcotic. Glycerine diethyl ether, which is only slightly soluble in water, but miscible in oil in all proportions, is of moderate strength, and glycerine triethyl ether, $C_3H_5(OC_2H_5)_3$, which is only faintly soluble in water, is a very potent narcotic (see also the phenols and their ethers, p. 133).

The acid esters of glycerine also have a narcotic effect. Once again, the monoester is weaker than the diester, and the diester is weaker than the triester, so long as the same acid moiety is involved. Triacetin, for example, produces narcosis at 1 : 200 (0.0009 mol/l).*

Tripalmitine, tristearine, and related compounds do not have a narcotic effect. On the one hand, they are of such extremely poor solubility in water that, even if they were readily soluble in lipoids of the ganglia cells, it would require weeks for equilibrium to be established between the aqueous solution of these fats and their level in the ganglia cells. Secondly, everything indicates that tripalmitine, tristearine, and related compounds are of relatively low solubility in swelled lecithin–cholesterol. Lecithin is barely soluble in olive oil; in the cold, cholesterol is soluble to barely 1%.

Acetal (ethylidene diethyl ether), $CH_3CH(OC_2H_5)_2$, is an excellent narcotic for tadpoles. Acetal is miscible with oil in all proportions, and soluble in water to approximately 4%. The olive oil/water partition coefficient for fairly diluted solutions is about 8.

Ethyl acetoacetate (acetoacetic ester) $CH_3COCH_2CO_2C_2H_5$, which also finds a place here, produces complete narcosis in about 1 : 400 = 0.019 mol/l, but rapidly causes death, no doubt as a result of a partial saponification. Death also follows within 5–6 hours in solutions of

* *Cf.* also the similar value of Baum, *loc cit.*, who also provides values for the monoacetins and the triacetins.

1 : 1000, which do not cause narcosis, but only sluggishness of movement. Acetoacetic ester is soluble in 13 parts water and 3.5 parts olive oil.

5.9 ACID AMIDES; UREA AND ITS DERIVATIVES

The two lowest members of the amides of monobasic fatty acids can hardly be considered narcotics. Acetamide does have a certain narcotic effect, but complete narcosis cannot be achieved with it, at least, not one that lasts for more than a few minutes without being fatal. In 1% solutions of acetamide (0.17 mol/l), tadpoles moved fairly normally for more than 18 hours, with weak clonic spasms occurring only from time to time. Tadpoles tend to die in 30–40 hours in 1% solutions without ever really becoming narcotized. The spasms are not caused by a hydrolysis of the acetamide into acetic acid and ammonia, since such cleavage takes place extremely slowly in plant cells and cold-blooded animals and the resulting free ammonia immediately passes through the epithelia of the gills to the outside. If a number of tadpoles are maintained in a small amount of acetamide solution in an enclosd container, and spirogyra, which contain tannic acid, are added, no precipitation of the cell fluid of the spirogyra occurs, even though a solution of 1 : 500 000 of ammonia produces a precipitate. A solution of 1 : 500 000 is, however, much too dilute to produce spasms in tadpoles. In 1–2% solutions, acetamide is harmful to most plant cells, also without causing actual narcosis. Acetamide penetrates the tissue more slowly than monohydric alcohols, but in smaller tadpoles, equilibrium between the level in the external solution and the tissues is essentially reached in about 2 hours. Acetamide is very readily soluble in water (in less than 0.5 parts), but is somewhat difficult to dissolve in oil (in 200 parts).

Amides of the monobasic fatty acids of higher molecular weight increase in narcotic strength only very gradually, according to the circumstance that in this series, the oil/water partition coefficient shifts only very gradually in favour of the oil. For example, valeramide, which is soluble in 9 parts of water, but only in 130 parts oil, produces complete narcosis only at about 0.5%.

Succinimide, $(CH_2CO)_2NH$, which is also much more soluble in water than in ether or olive oil, has hardly any narcotic strength. In 2% solutions (0.2 mol/l), tadpoles, for example, continued to move around for more than 24 hours, but died within 2–3 days. Even in a 4% solution, the movements after 6 hours were only sluggish, but hardly disappeared. Succinimide penetrates the tissue about as quickly as acetamide.

The behaviour of urea and thiourea and their derivatives is very interesting. Urea and thiourea themselves do not have the slightest narcotic effect. Both also penetrate the blood and tissue only very slowly, so that 24, 36 or more hours are required to reach equilibrium with respect to the concentration of urea in the external solution and the tissues. Solutions of urea of more than 1.25% accordingly have a dehydrating effect if tadpoles are placed in them directly. If, however, the concentration is increased gradually, the tadpoles can tolerate solutions of 1.25% or more without the slightest disturbances occurring, with their movements remaining very lively. Urea dissolves in about one part cold water, but is only slightly soluble in ether and oil. Thiourea is also very soluble in water, but only very slightly soluble in ether and oil.

Methylurea, $CO(NH_2)NHCH_3$, and methylthiourea, $CS(NH_2)NHCH_3$, which are somewhat less soluble in water and somewhat more soluble in ether than the unsubstituted compounds, also do not yet have a narcotic effect. However, the compounds phenylurea, $CO(NH_2)NHC_6H_5$, and phenylthiourea, $CS(NH_2)NHC_6H_5$, do act as narcotics, corresponding to the fact that the addition to a molecule of one alkyl group with many carbon atoms in place of a hydrogen atom produces a greater shift in the partition coefficient between water and an organic solvent than the replacement of a hydrogen atom by a methyl group.

Phenylurea, which also penetrates the tissue much more rapidly than urea, produces complete narcosis in tadpoles within about 30 minutes at about 1 : 1600 (0.0046 mol/l). The narcosis can last 20–24 hours without causing death. Although admittedly the circulation was almost gone, tadpoles lived for 12 hours in a 1 : 800 solution, and recovered completely within several hours after transfer to fresh water. Phenylurea dissolves to about 0.5% in cold water, and in 100 parts of ether.

Phenylthiourea produces narcosis at approximately 1 : 1000 (0.0066 mol/l). It is somewhat less soluble in water than phenylurea.

Phenylacetylurea is transported much more rapidly through the epithelia of the gills and skin into the blood and from there into the cell tissues than urea, but much more slowly than the monohydric alcohols. In 2% solutions (0.172 mol/l), hardly any effect can be seen in the first 20 minutes. After 30–40 minutes, the tadpoles become somewhat sluggish in their movements. At this moment, the circulation is still fairly good, but soon becomes weaker. Circulation is extinguished after 2–3 hours, though the response to stimuli remains more or less intact until death.

Triethylthiourea penetrates the blood and tissues rapidly and is a good, though not very potent narcotic. It induces complete narcosis only at about 0.4%. This compound dissolves in approximately 60 parts of water and 10 parts of oil.

5.10 CHLORALOSE

Chloralose $C_8H_{11}Cl_3O_6$ occupies a somewhat peculiar position among the non-specific narcotics. The constitution of this compound has not yet been sufficiently elucidated*. However, since it contains one tetra-acetate and one tetrabenzoate, it must contain four hydroxyls. Strangely enough, however, chloralose is more soluble in ether than in cold water, a behaviour that is surprising in a compound with so many hydroxyls. Chloralose dissolves in about 150 parts of cold water and in about 1600 parts of oil.

The phenomena of narcosis with chloralose are not very easy to interpret, and can only be fully explained after further studies have been made. If tadpoles are placed in a 0.5% solution (0.016 mol/l), their movements are hardly affected for the first 30 minutes. There is, however, a distinct tendency for clonic spasms. After 1 hour and 30 minutes, the tadpoles become almost immobile, but respond well to stimuli. After 2 hours, the circulation is still good, with 70 fairly strong pulse beats per minute. In one experiment, after 3 hours 45 minutes, one tadpole was still slightly responsive, but two others were already completely unresponsive, although the circulation was still retained. The circulation ceased after 6–8 hours, and the tadpoles did not recover when they were transferred to fresh water.

In 0.25% solutions (0.008 mol/l), 15–18 mm long tadpoles moved around almost normally for about 1 hour, although slight spasms occasionally occurred. After 2 hours, they mostly moved only when stimulated. After 4 hours, they were still very responsive to stimuli, but did not move otherwise. After 6 hours and 30 minutes, their responsiveness ceased, and their circulation was rather weak. When they were subsequently transferred to fresh water, the tadpoles did not recover, and death followed within a few hours.

In 0.1% solutions (0.0032 mol/l) of chloralose, tadpoles moved around almost normally for 2–2.5 hours, with no sign of spasms. After 3 hours and 30 minutes, only a few spontaneous, sluggish movements were occurring. After 5 hours, the tadpoles moved only in response to stimuli, and even then, only for 1–2 seconds. After 12 hours, responsiveness had still not quite ceased. After 20 hours from the beginning of the experiment, death occurred.

These test results can be partially explained by the fact that chloralose penetrates the tissue rather slowly, and passes out equally slowly. This accounts for the very slow onset of narcosis, even in 0.5% solutions. Also, the phenomenon that tadpoles survived in a 0.25% solution for a fairly

* Editor's note: the chemical structure for α-chloralose is (R)-1,2-*O*-(2,2,2-trichloroethylidene)-α-*D*-glucofuranose (Taga, T. *et al.* (1982) *Acta Crystallogr.*, **B38**, 1874).

long time but did not recover after transfer to fresh water, even though the circulation had not stopped, can probably be explained as follows: chloralose passes slowly out of the tissues and the blood into the surrounding water, and when a compound acts on the heart, it is the length of time that it acts, and not just its concentration in the blood, heart musculature, and ganglia cells related to the heart, that is of great importance. This has been demonstrated most clearly by experiments with ether solutions of varying concentrations. It can be proven by means of plasmolytic experiments that chloralose actually penetrates living cells relatively slowly. It is, incidentally, the only non-specific narcotic known to me that possesses this characteristic of slow penetration.

Parachloralose, which is only very slightly soluble in most solutions, has absolutely no narcotic effect. Tadpoles can live for weeks in its saturated aqueous solution.

REFERENCES

1. Binz, C. (1891) *Vorlesungen über Pharmakologie für Artze und Studirende*, 2nd edn, A. Hirschwald, Berlin, p. 177.
2. Binz, C. *Vorlesungen über Pharmakologie für Artze und Studirende*, 2nd edn, A. Hirschwald, Berlin, p. 178 and (1880) *Arch. exp. Pathol. und Pharmakol.*, **13**, 139.
3. Arloing, S. (1879) *Recherches expérimentales comparatives sur l'action du Chloral, du Chloroforme et de l'Ether avec applications practiques*, Paris, Libraire de l'Academie de Médecin. 178pp.
4. See Reicher, *Lieb. Ann.* 1885, **128**, p. 257, and especially Loewenherz: *Zeitschr. physik Chem.* 1894, **15**, p. 389; further, the handbooks and textbooks on theoretical chemistry by Ostwald and Nernst; as well as the lectures on theoretical and physical chemistry of van't Hoff.

6

Aromatic compounds

Until very recently, the statement has appeared in the literature that no actual non-specific narcotic is known among the aromatic compounds. In fact, as long ago as 1848, no less a person than Simpson found that benzene itself is a narcotic, although it proved to be unsuitable for medical use. More recently, Meyer has studied the narcotic effect of the amides of the various aromatic acids. We will see in the following that among the aromatic compounds can be found some of the most interesting non-specific narcotics, and that it is within this group that the relationships among the non-specific narcotics, antipyretics, and antiseptics on the one hand, and between non-specific and basic narcotics on the other, can be most clearly seen. Without further introduction, I will proceed to a discussion of the specific groups.

6.1 AROMATIC HYDROCARBONS AND AZOBENZENE (Table 6.1)

All of these compounds penetrate the blood and tissues very readily. In solutions of 1 : 4000 benzene, for example, 16–17 mm long tadpoles become fully narcotized within 2½ minutes. In contrast, at 1 : 8000, complete narcosis does not take place even within 24 hours. Narcosis induced by benzene is fatal within a few hours. Strong agitation occurs before the onset of narcosis and this can produce spasms. In solutions of naphthalene,

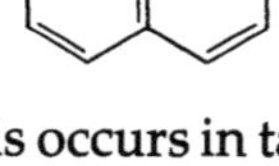,

of 1 : 50 000, complete narcosis occurs in tadpoles 11–12 mm long after 10 minutes. After 6 minutes, it is not quite complete. In 1 : 200 000 solutions

Table 6.1 The aromatic hydrocarbons and azobenzene

Compound	*Concentration for narcosis in tadpoles*		*Solubility*
	(parts by weight of water per one part by weight of narcotic)	*(mol/l)*	
Benzene	6000	0.0021	Soluble in about 1000 parts of water; miscible with oil
Xylene	25 000	0.00038	Soluble in about 8000 parts of water; miscible with oil
Naphthalene	100 000–150 000	0.000052–0.000078	Soluble in about 20 000–30 000 parts of water; readily soluble in ether and oil
Phenanthrene ($C_{14}H_{10}$)	1 500 000	0.0000037	Soluble in about 300 000 parts of water; readily soluble in ether and oil (14%)
Anthracene ($C_{14}H_{10}$)	The supersaturated aqueous solution (1 : 1 000 000) does not narcotize; after 14 days, the movements are hardly slower than under normal conditions	—	Soluble only in excess of 1 000 000 parts of water; also very difficult to dissolve in cold ether and cold olive oil
Azobenzene $(C_6H_5)_2N_2$	300 000	0.000018	Soluble in about 20 000 parts of water; very readily soluble in olive oil

of naphthalene, complete narcosis does not occur even after several days. In 1:50 000 azobenzene, $(C_6H_5)N_2$, 12 mm long tadpoles still moved somewhat after 8 minutes, but became completely unresponsive after 14 minutes, although the circulation was still good after 22 minutes. In 1:200 000 azobenzene, 12 mm long tadpoles still moved around fairly vigorously after 1 hour and 30 minutes. After 2 hours 45 minutes, they did not move around, but responded readily to stimuli. When stimulated, they even moved for about 3 seconds. Narcosis was complete after 50 hours and the tadpoles did not respond to even the strongest stimuli, although circulation was still excellent. When the concentration was then decreased to 1 : 400 000, narcosis was maintained for about 16 hours while the tadpoles became somewhat responsive to stimuli, but remained immobile. The circulation was good. When placed directly in a solution of 1:400 000 azobenzene, tadpoles become almost completely immobile after a few hours, but remain responsive to stimuli for days.

6.1.1 Potent narcotic action of phenanthrene

In 1 :1 000 000 phenanthrene, narcosis was almost complete after 6 hours, but deepened very slowly. After 9 hours, there was still a faint responsiveness but the last trace of this disappeared after 11 hours. After 34 hours from the beginning of the experiment, or 23 hours following the onset of complete narcosis, the circulation was still present, but very sluggish. When the tadpoles were transferred to fresh water, the circulation was almost entirely normal again after another 15 hours, but the tadpoles hardly responded to stimulus. The tadpoles recovered completely after another 24 hours. In a 1 : 1 500 000 solution, 36 hours were needed for complete narcosis to take place. In a 1 : 2 000 000 solution, tadpoles did not become fully narcotized even after 10 days, but only become sluggish.

One can see from the time frames given here that, in the case of these very potent narcotics, the rather slow onset of narcosis is not caused by the fact that the epithelia of the gills and the ganglia cells are not very permeable to the narcotics, but simply by the fact that, in each circulation of the body, even if there is almost complete equilibrium between the concentrations of the blood plasma and the external solution, the blood can transport only a small quantity of the narcotic to the ganglia cells. Also, because of the extraordinary size of the (brain lipoid/water) partition coefficient, the ganglia cells can become saturated with the narcotic only very gradually. The extremely small absolute difference in narcotic concentration between the external liquid and the blood plasma and again between the blood plasma and the intercellular lymph, i.e. after equilibrium is 75–90% complete, also causes very slow diffusion of the narcotic, for the rate of diffusion does not depend on the relative differences in the concentrations, but on the absolute differences. In addition, a certain volume of blood can absorb a greater amount of the narcotic from the solution surrounding the gills and pass it to the ganglia cells than can the same volume of water or blood plasma, since the blood corpuscles accumulate the narcotic as a result of their lecithin–cholesterol content, and then return it partially to the blood plasma as soon as the concentration of the narcotic in the blood plasma drops. The lipoids of the blood corpuscles play a similar role in transporting these narcotics to that played by the haemoglobin in transporting oxygen, even though the laws governing the increase and decrease take a different form.

One very instructive fact is that, while phenanthrene is such an extraordinarily strong narcotic, the isomeric anthracene has no narcotic effect. Phenanthrene is very soluble in ether and oil (14%), but anthracene is very insoluble. In the case of phenanthrene, its maximum narcotic strength has also certainly not been reached. Retene, for example, is still readily soluble in ether and other such solvents. According to the general rules, however, its partitioning between water and ether must favour the organic solvent even more than in the case of phenanthrene. Fluoranthrene and pyrene may also turn out to be more powerful narcotics than phenanthrene. However, such large quantities of solution are needed, and the experiments with these compounds require so much time, that I did not want to extend my experiments to include these compounds. With phenanthrene, 1–2 litres of solution is required for the tests so that no significant change in concentration occurs.

In connection with this group of narcotics, it should still be mentioned that saturated, or somewhat supersaturated aqueous solutions of anthraquinone ($C_6H_4C_2O_2C_6H_4$) have just as little narcotic effect as anthracene. Like anthracene, anthraquinone is only slightly soluble in oil and water.

From the point of view of method, it should also be mentioned that it is best not to prepare directly the aqueous solutions of these compounds that are extremely difficult to dissolve in water (benzene and xylene are not included here). Instead, a solution of appropriate strength in ethyl alcohol should be prepared first and a certain amount of this alcoholic solution should be added dropwise to the water with vigorous stirring. I have found by numerous control experiments that less than 1 part per 1000 of ethyl alcohol in the aqueous solution has absolutely no effect on the test results.

6.2 PHENOLS AND THEIR ETHERS, VANILLIN AND PIPERONAL

The effects of phenols and their ethers are extraordinarily interesting. The 'neutral' ethers and the monovalent and polyvalent (i.e. in this case, with

Table 6.2 Effects of phenols and their ethers and of vanillin and piperonal

Compound and formula	*Most important effects at different concentrations*	*Solubility in water and olive oil, etc.*
Phenol (C_6H_5OH)	In solutions of 1 : 2000, narcotizing and paralysing on the circulation, with a tendency for spasms; action on the heart at this concentration; this action still strong at 1 : 4000	Soluble in water up to 7%; soluble up to 25% in oil
Anisole ($C_6H_5OCH_3$)	Purely narcotic effect; complete narcosis at 1 : 6000 (0.0015 M)	Soluble in 650 parts of water; miscible with oil
Cresols (ortho, meta and para) ($C_6H_4(CH_3)(OH)$)	Effects as with phenol; but narcotic effect somewhat more predominant; 0.0018 M (1 : 5000) leads to heart paralysis and death	Soluble in 50–250 parts of water; soluble in less than 1 part of oil (para), or miscible with oil (ortho) and meta
Amylphenol ($C_6H_4(C_5H_{11})(OH)$)	Predominantly narcotic effect; at first some tendency to spasms. Already less than 1 : 200 000 (0.000030 M) produces complete narcosis	—
Thymol ($C_6H_3(CH_3)(C_3H_7)(OH)$)	Effect (in very dilute solution) purely narcotic; complete narcosis at 1 : 120 000 (0.000055 M)	Soluble in 1200 parts of water and in 1 part of oil (at 20°C)
Naphthol (α and β) ($C_{10}H_7OH$)	Like phenol, the effect is predominantly on the heart; tendency for spasms. Narcotic effect very obvious. Heart paralysis a few hours at 1 : 80 000 (α-napthol)	—
Resorcinol ($C_6H_4(OH)_2(1,3)$)	Effect primarily on the heart, with tendency for spasms. Almost no narcotic effect	Soluble in less than 1 part of water; soluble in about 8–10 parts of oil (at 20°C)

Resorcinol dimethyl ether ($C_6H_4(OCH_3)_2(1,3)$)	Purely narcotic effect. Complete narcosis at 1 : 16 000 (0.00045 M). The narcosis can be sustained for a long time without producing death	Soluble in about 900–950 parts of water; miscible with oil (at 18°C)
Hydroquinone ($C_6H_4(OH)_2(1,4)$)	Effect like resorcinol, but stronger	—
Hydroquinone dimethyl ether ($C_6H_4(OCH_3)_2(1,4)$)	Purely narcotic effect. Complete narcosis at 1 : 8000 (0.0009 M)	Soluble in 1300 parts of water and in 3.5 parts of oil (at 18°C)
Pyrocatechol $C_6H_4(OH)_2(1,2)$	Effect like resorcinol, but stronger	—
Guaiacol ($C_6H_4(OCH_3)(OH)(1,2)$)	Effect like the cresols	Dissolves in 50 parts of water and in less than 0.5 parts of oil (at 20°C)
Orcinol ($C_6H_3(CH_3)(OH)_2(1,3,5)$)	Effect exactly like resorcinol	Dissolves in 5 parts of water and in about 14 parts of oil (20°C)
Eugenol ($C_6H_3(CH_2CHCH_2)$-$(OCH_3)(OH)(1,3,4)$)	Effect predominantly narcotic, but after a while, also paralysis of the heart. Complete narcosis at 1 : 50 000 (0.00012 M)	Dissolves in about 1000 parts of water; miscible with oil
Phloroglucinol ($C_6H_3(OH)_3(1,3,5)$)	Effect very weak and indistinct; gradual paralysis of the heart at higher concentrations	Dissolves in about 60 parts of water and in 2500 parts of oil (at 25°C)
Pyrogallol ($C_6H_3(OH)_3(1,2,3)$)	Effect much stronger than that of phloroglucinol; indeed also paralyses the heart	—
Vanillin ($C_6H_3(CHO)(OCH_3)(OH)(1,3,4)$)	Effect is at first narcotic, later also paralysing of the heart. Complete narcosis at 1 : 2000 (0.0033 M)	Dissolves in about 100 parts of water and in 35–40 parts of oil (at 18°C)
Piperonal ($C_6H_3(CHO)(OCH_2O)(1,3,4)$)	Effect is at first narcotic. Complete narcosis at 1 : 2000 (0.0033 M). Solutions of 1 : 4000 do not produce complete narcosis, but nevertheless gradually paralyse the heart	Dissolves in about 550 parts of water and 6 parts of oil (at 18°C)

more than the phenol hydroxyl) phenols are all good and strong narcotics that leave the circulation intact for a long time. For the most part, monovalent phenols produce spasms and cessation of heartbeat, along with narcosis, although some phenols have an effect that is almost purely narcotic. The dihydroxybenzenes and trihydroxybenzenes have a very strong effect upon the heart (with the exception of phloroglucinol, which has little effect), while the narcotizing effect on the brain is quite insignificant. Their 'acid ethers' with one phenol hydroxyl, act in substantially the same way as the monovalent phenols. The neutral phenol ethers and all monovalent phenols penetrate very rapidly into the blood and the cell tissues and pass out of them equally quickly when the gill-breathing test organisms are transferred to fresh water. The dihydroxybenzenes penetrate somewhat more slowly, though still quite rapidly, and the trihydroxybenzenes are somewhat slower still.

I will provide in the following sections, in as condensed a format as possible, the most important test results for the individual compounds listed in Table 6.2.

6.2.1 Monovalent phenols and their ethers

Phenol (carbolic acid)

In a 1 : 2000 (0.0053 mol/l) solution of phenol, the movements of 15–16 mm long tadpoles are very srongly affected. Long periods of quiet alternate with short jerky movements. Later, the tadpoles move only when stimulated, with quivering spasms occurring from time to time. Circulation ceases or has almost ceased after 2 hours, but there is still response to stimulus. If the tadpoles are then transferred to fresh water, almost immediately (in 1–2 minutes) they become more mobile, and after 5 minutes, their movements are almost normal, although the circulation remains imperceptible for a long time. After another 2 hours, the circulation of some of the tadpoles begins to be restored and after 4 hours, it is essentially normal in the surviving tadpoles. Usually, however, a large percentage of the tadpoles have died before this occurs.

In a 1 : 4000 (0.0026 mol/l) solution of phenol, the movements of the tadpoles take on a somewhat spasmodic character in the first few minutes. After 1–2 hours, circulation is slow and the blood pressure had dropped considerably, so that the blood does not flow into all of the capillaries. This condition can last for another 36 hours without much deterioration and the tadpoles continue to exhibit a spontaneous jerky movement throughout the entire period. The ganglia cells of the cerebrum are relatively little affected.

In any case, the fairly extensive narcosis in stronger phenol solutions (1 : 2000) must be independent of circulatory disturbance, as can be seen from the fact that narcosis fades immediately when the test organisms are transferred to fresh water, while the circulation does not improve for a fair amount of time. The same thing can be seen when a comparison is made between the effects of phenol and dihydroxybenzenes. Phenol is soluble in water to about 7% and in olive oil to about 25% (much more when heated). Narcosis in the ganglia cells is probably only partly the result of phenol entering the brain lipoids. It also partly results from a loose (in a partial state of dissociation) bond between phenol and the proteins of the ganglia cells, as comparative studies appear to have shown. In solutions of 1 : 2000 phenol, which exerts only a strongly narcotic effect on the ganglia cells of tadpoles, most plant cells and animal cells of lower order are also affected, but the effect is like that of the dihydroxybenzenes, which act mainly on the cell proteins. Furthermore, the oil/water partition coefficient, is too small to account for the relatively strong narcotic effect of phenol if only the brain lipoids are affected. It is also known that carbolic acid forms an actual bond with proteins.

Cresols, $C_6H_4(OH)CH_3$

All three cresols have effects similar to those of phenol, but their effects take place at somewhat lower concentrations. Their narcotic effect, however, is somewhat more dominant than with phenol, though it is less than the effect on the circulation.

In solutions of 1 : 7500 (0.0012 mol/l) orthocresol, tadpoles still respond when stimulated after 1 hour and 30 minutes, but are otherwise immobile. After this amount of time, no circulation is detectable in the tail, and the tadpoles become sluggish in the other parts of the body. After 4 hours, the tadpoles are quite unresponsive to stimulus, and the circulation has ceased completely. Nevertheless, when the tadpoles are returned to fresh water, their responsiveness returns within 10 minutes. The circulation, however, does not return to normal for several hours. In solutions of 1 : 10 000 (0.0009 mol/l), tadpoles retain their ability to move around for about 36 hours, but their movements are jerky and convulsive, and their circulation becomes extremely sluggish.

In solutions of 1 : 5000 (0.0018 mol/l) metacresol, tadpoles remained responsive to stimulus for more than 1 hour, but their circulation became extremely sluggish, with death following after about 4 hours. In solutions of 1 : 10 000 (0.0009 mol/l), tadpoles remained somewhat mobile for about 36 hours, but their movements were jerky and convulsive, with circulation very sluggish.

Paracresol has a somewhat weaker effect than orthocresol and metacresol. In solutions of 1 : 5000, tadpoles do not lose their responsiveness to stimuli for 3–4 hours. In 1 : 10 000, it results in fatality more quickly than with the other two cresols at the same concentration.

About 2.5 parts of orthocresol, 0.5 parts metacresol, and 1.0 parts of paracresol dissolve in 100 parts of water. All three cresols are very soluble in olive oil. The mechanism of action of cresol is undoubtedly, like that of phenol, a combined one, in that on the one hand cresols accumulate in brain lipoids, and on the other hand a partial reaction occurs between the cresols and the proteins of the cells.

The important antiseptic lysol, which for the most part consists of a solution of various cresols in linseed oil–potash soap, in which undoubtedly loose or partially dissociated bonds with the cresols exist, owes its antiseptic properties to the free cresols. These dissociate to almost the same degree that they are absorbed by the protoplasm, bacteria, and other cells, so that the concentration of the active lysol constituents remains nearly constant for a long time. Cresols in purely aqueous solution can sometimes undergo a large decrease in concentration as a result of their accumulation by fatty substances, and in the absence of liquid turbulence, this decrease is equalized only slowly by diffusion.

Amylphenol, $C_6H_4(OH)C_5H_{11}$

This compound was tested under the assumption that, because of the magnitude of its oil/water partition coefficient, its narcotic effect would be so insignificant that the other effects of the compound, which result simply from its containing one phenolic hydroxyl, would also disappear. However, this assumption has been only partly confirmed, in that it obviously retains some of the effects of phenols of lower molecular weight. In solutions of 1 : 100 000 (6.1×10^{-5} mol/l), spasms occur in tadpoles in the first 2 minutes similar to those produced by carbolic acid solutions. After about 4 minutes, the tadpoles become immobile, even though the circulation is still quite good after 20 minutes. In 1 : 200 000 (3.0×10^{-5} mol/l), there is at first a tendency to spasms. After about 30 minutes the tadpoles become unresponsive to stimulus, although the circulation remains good. After 9–10 hours, or even sooner, circulation ceases. The relatively rapid occurrence of narcosis in such a diluted solution as 1 : 200 000 makes it probable that complete narcosis would gradually take place in even more dilute solutions. Furthermore, it is likely that the fairly rapid cessation of heart beat in this 1 : 200 000 solution only corresponds to the general rule that the non-specific narcotics quickly produce paralysis of the heart when their concentration in the blood is significantly elevated above that just sufficient for narcosis.

Thymol, $C_6H_3(CH_3)(C_3H_7)(OH)(1,2,3)$

In sufficiently diluted solution, thymol has a purely narcotic effect on tadpoles without causing spasms or circulatory disturbances. At

1 : 30 000, 15 mm long tadpoles become paralysed within 2 minutes. After 5 minutes, they do not respond to stimulus. In 1 : 60 000, their responsiveness is still retained after 15 minutes, but the tadpoles do not move spontaneously, and after 30 minutes, they are fully narcotized. After 3 hours in these solutions, the circulation is still good. In solutions of 1 : 120 000 (5.5×10^{-5} mol/l), tadpoles remain somewhat responsive for hours, but do not make any spontaneous movements after 30 minutes. After 16–20 hours, most of them are completely unresponsive, although the circulation is still good. After being kept in the solution for 40 hours, then transferred to water, most of the tadpoles recover completely within 1–2 hours. Thymol dissolves in approximately 1200 parts water and one part olive oil.

α-Napthol and β-napthol
These act substantially like carbolic acid, but in more diluted solutions. For example, in 1 : 80 000 α-napthol, tadpoles remain responsive to stimulus for several hours, but the circulation soon becomes sluggish, and their movements have a spasmodic character. At these concentrations, the solution is fatal in 12–16 hours. β-napthol acts similarly, but is somewhat weaker.

Anisole, $C_6H_5OCH_3$
In contrast to phenol itself, the ethers of phenol have a purely narcotic effect without causing spasms or affecting the heart any further. In 1 : 4000 (0.0023 mol/l), anisole produces complete narcosis in tadpoles within three minutes. At 1 : 6000 (0.0015 mol/l), narcosis takes place within 6–8 minutes, and can be maintained for more than 20 hours without being fatal. In 1 : 8000, tadpoles become very sluggish, but never fully narcotized. They can survive for days in this solution.

6.2.2 Divalent phenols and their 'acid' and neutral ethers

Resorcinol
In solutions of 1 : 400 (0.023 mol/l), tadpoles remain more or less mobile for more than 30 minutes. During the first 5 minutes, their movements are almost normal, although there is a tendency to weak clonic spasms. Later, only isolated jumpy movements occur after long periods of quiet, as in solutions of phenol. After 30–40 minutes, circulation ceases completely, although isolated spontaneous body jerks can be observed for a longer period of time. If the tadpoles are transferred to fresh water within 1 hour, they usually recover completely after a few hours. In solutions of 1 : 800, circulation is maintained for 1 hour, though it becomes very

weak, while the movements of the tadpoles remain rather lively, but short and jumpy. In solutions of 1 : 2000, tadpoles survive for about 24 hours without becoming narcotized. In any case, resorcinol exerts only a slight effect on the sensory system, which is much less than on the heart.

Pyrocatechol and hydroquinone act similarly to resorcinol, but in solutions about three times more dilute. The epithelia are also heavily affected in solutions of 1 : 2000.

It is noteworthy that these three dihydroxybenzenes exert a very obvious effect on plant cells at the same concentrations that are required to produce more noticeable symptoms in tadpoles. In 1 : 400 resorcinol, for example, plant cells become narcotized in a few minutes. The narcosis has a completely different character from narcosis induced by ether, chloroform, and other true non-specific narcotics, in that the protoplasm assumes a certain rigidity, which disappears again in a few minutes upon transfer to fresh water. I immediately noticed this effect of resorcinol and other hydroxybenzenes on the protoplasm when I did my first experiments eight years ago with these compounds. The whole course of these phenomena led me to the conclusion that these hydroxybenzenes form a loose bond with the proteins in the cell which is reversed again when the hydroxybenzenes are removed from the solution, so long as the effect is not too great, which would, of course prove fatal to the cells.

All of these dihydroxybenzenes are more soluble in water than in olive oil.

Resorcinol dimethyl ether, $C_6H_4(OCH_3)_2$

This compound is likewise an excellent narcotic for tadpoles and other organisms that respire by means of gills. In solutions of 1 : 8000 (9×10^{-4} mol/l), 10–12 mm long tadpoles become fully narcotized in less than 2 minutes. After 10 hours, the circulation becomes very sluggish, but the tadpoles recover almost completely within 8–10 minutes after transfer to fresh water, while the circulation also quickly improves. In 1 : 16 000 (4.5×10^{-4} mol/l) complete, or nearly complete narcosis is produced within 5–6 minutes, and the tadpoles recover in a few minutes when transferred to fresh water after a narcosis of more than 60 hours in this solution. During this entire period of narcosis, the circulation remains good. Resorcinol dimethyl ether dissolves in about 950 parts water and is miscible with ether in all proportions.

Dimethyl hydroquinone $C_6H_4(OCH_3)_2$

This acts quite similarly to resorcinol dimethyl ether, but has a somewhat lower narcotic strength. Solutions of 1 : 10 000 do not narcotize completely; solutions of 1 : 8000 (9×10^{-4} mol/l) produce complete narcosis. With this compound, narcosis can also be maintained for a very long time without being fatal or weakening the circulation to a great extent. Hydroquinone dimethyl ether dissolves in about 1300 parts water and in 3.5 parts olive oil at 18°C.

Orcinol, $C_6H_3(CH_3)(OH)_2(1,3,5)$

This acts quite similarly to resorcinol, both qualitatively and quantitatively. It affects the heart more strongly than the sensory system. A kind of narcosis occurs in stronger solutions, but it seems to be mainly the result of circulatory disturbance, since after transfer to fresh water, responsiveness returns only very slowly, and at the same rate at first with the improvement in circulation. Orcinol is more soluble in water than in olive oil.

Guaiacol, $C_6H_4(OCH_3)(OH)(1,2)$

This acts similarly to cresols, though it requires somewhat higher concentrations. In solutions of 1 : 3000 (3.1×10^{-3} mol/l), tadpoles move about for some 24–30 hours, but only in fits and starts after long quiet periods, with the circulation soon becoming very weak. In solutions of 1 : 2000, they die within a few hours, and in this case, the circulation ceases before a lack of response occurs. Guaiacol dissolves in about 50 parts of water, and in less than 1 part of oil.

Eugenol, $C_6H_3(CH_2CH{=}CH_2)(OCH_3)(OH)(1,3,4)$

This is a much better narcotic, but it has a rather strong effect on the circulation. In 1 : 50 000 (1.2×10^{-4} mol/l), tadpoles are completely narcotized. The circulation is weakened at first, but ceases completely after about 4 hours. In solutions of 1 : 75 000, circulation is maintained for about 36 hours, but the narcosis in these solutions is incomplete. Eugenol is very slightly soluble in water (about 1 part per 1000), but very soluble in olive oil.

Phloroglucinol, $C_6H_3(OH)_3(1,3,5)$

This has only a slight effect on tadpoles, and it also has much less effect on plant cells than the dihydroxybenzenes. In 0.5% solutions (0.04 mol/l), within the first hour tadpoles behave almost the same as in fresh water. Their movements gradually become somewhat affected, but after 24 hours, they are still rather lively. The circulation becomes sluggish after 24 hours. In 0.25% solutions (0.02 mol/l), tadpoles remained vigorous for over 80 hours. After 96 hours, they had become somewhat sluggish and the circulation was still rather good, but slow. Within an additional 24 hours, the solution was fatal to the tadpoles.

In solutions of 1 : 1000 (8×10^{-3} mol/l), tadpoles were just as lively and vigorous after 10 days as at the beginning of the experiment.

Phloroglucinol dissolves in about 60 parts of water and in 2500 parts of olive oil.

Pyrogallol, $C_6H_3(OH)_3(1,2,3)$
This has a much stronger effect than phloroglucinol, but its effect occurs more slowly. In a 0.5% (0.04 mol/l) solution, 16–17 mm tadpoles move almost normally within the first 15–20 minutes, but their circulation is already very weakened, and stops completely a few minutes later, although the mobility of the tadpoles is at first retained. If the tadpoles are maintained in the solution for more than 25 minutes, they do not recover after transfer to fresh water, even if they are still rather lively at the time of transfer. In solutions of 1 : 1000 (8×10^{-3} mol/l) within about an hour circulation ceases entirely in the tails of the tadpoles, and they become extremely sluggish in other parts of the body. The tadpoles fail to respond to stimuli after 2–3 hours.

Both phloroglucinol and pyrogallol penetrate the epithelia of the gills into the bloodstream, and from there into the tissues somewhat more slowly than the monovalent or divalent phenols, thus accounting for their slight effect during the first few minutes. Pyrogallol and the other dihydroxybenzenes probably exert, in addition to their other effects, a reducing action of the cell tissues, or at least consume a portion of the free oxygen dissolved in the cell fluids.

6.2.3 Vanillin and piperonal

In connection with the hydroxybenzenes and the ethers, the effects of vanillin and piperonal will be mentioned briefly.

Vanillin, $C_6H_3(CHO)(OCH_3)(OH)(1,3,4)$
In solutions of 1 : 2000 (3.3×10^{-3} mol/l) of vanillin, tadpoles become fully narcotized within 10–15 minutes. If they are kept for only about 1 hour in the solution, they recover completely within 15 minutes. In solutions of 1 : 3000, their response to stimuli did not disappear until after more than an hour. After 6 hours and 30 minutes in this solution, the tadpoles only partially recovered, and subsequently died. In solutions of 1 : 4000, the circulation was always found to suffer after a while.

Vanillin dissolves in approximately 100 parts of cold water and in 35–40 parts of oil.

Piperonal, $\underbrace{C_6H_3(CHO)OCH_2O}(1,3,4)$
In solutions of 1:1000 (6.7×10^{-3} mol/l) of piperonal, tadpoles become completely narcotized in 2–3 minutes and recover in about 15 minutes

after an exposure of 20–30 minutes. In a 1 : 2000 solution, complete narcosis occurs within 5–10 minutes; however, if the narcosis is maintained for more than about 4 hours, the tadpoles do not recover. In 1:4000, tadpoles remain responsive to stimulus for more than 3 hours. However, within this period, the circulation has already become very weak, and this concentration is fatal within another 3–4 hours. In solutions of 1:8000, tadpoles remained lively for more than two weeks. Piperonal dissolves in about 550 parts water and about 6 parts oil.

Table 6.2 gives the most important data on the effects of hydroxybenzenes and their ethers and of piperonal and vanillin. The order in which the compounds are presented differs somewhat from the text. With the assistance of Table 6.2, the effect of the aromatic hydroxyl group can readily be surveyed, as well as the decrease in this effect in those compounds containing greater numbers of carbon and its disappearance when the hydroxyl hydrogen is replaced by alkyl substituents.

6.3 OIL OF TURPENTINE, CAMPHOR AND VOLATILE OILS

Turpentine, $C_{10}H_{16}$

In the fresh or non-oxidized state, turpentine acts more or less as a pure narcotic. Complete narcosis occurs at approximately 1 : 15 000 (5×10^{-4} mol/l). The presence of some oxidized turpentine affects the epithelia.

Terpin hydrate, $C_{10}H_{20}O_2 \cdot H_2O$

This has absolutely no effect on tadpoles, even in a 1 : 250 saturated solution. They remained lively and vigorous for weeks in a saturated solution at this concentration. Terpin hydrate dissolves in 100 parts of cold ether and is also of low solubility, i.e. it dissolves in approximately 700 parts of cold olive oil.

Menthol, $C_{10}H_{19}OH$

This compound is the hexahydro- derivative of thymol, and has the character more of an aliphatic alcohol than of a phenol. In solutions of 1 : 10 000, 12–13 mm tadpoles of *Rana temporaria* become completely narcotized within 80 seconds. In solutions of 1 : 20 000, narcosis is produced in 3–4 minutes, while at 1 : 40 000, 10–12 minutes are required. Complete narcosis can be achieved after some time in a 1 : 60 000 (1.1×10^{-4} mol/l) concentration, but no longer at 1 : 80 000. Menthol dissolves in approximately 1500 parts of water, and in 5 parts of olive oil.

Camphor, $C_{10}H_{16}O$

In more dilute solutions, for example 1 : 40 000 or 1 : 20 000, of normal (Japan) camphor, which is an alicyclic ketone, tadpoles display a considerable degree of agitation. This agitated state persists for a very long period of time without the tadpoles exhibiting any further signs of abnormality. They can survive for days in solutions of 1 : 12 000, without their movements being noticeably affected. In solutions of 1 : 5000 (1.3×10^{-3} mol/l), all spontaneous movements cease after about 15 minutes, although the tadpoles continue to respond to stimuli for about 1 hour. After complete narcosis has occurred, the tadpoles survive for about 6 hours, but their circulation becomes very slow. The tadpoles do not usually recover after being maintained in the solution for 8–10 hours.

Aqueous solutions of all essential oils have a rather marked narcotic effect, but many also have additional effects at the same time, e.g. paralysis of the heart. In very dilute solutions, they also frequently show a camphor-like agitating effect. These solutions also often affect the epithelia, which in many cases probably reflects the level of peroxide present. Although I have performed numerous experiments on the narcotic effects of these oils, mainly in the interest of certain problems in plant physiology, it seems unnecessary to explore these in further detail. It will suffice to say that narcosis usually occurs in solutions between 1 : 10 000 and 1 : 100 000 and that they are fatal within a few minutes to several hours, depending upon the essential oil used.

6.4 LACTONES AND ANHYDRIDES

Among these compounds, only phthalide and coumarin were carefully tested.

Phthalide, $\underbrace{C_6H_4COOCH_2}$

In solutions of 1 : 1500 of phthalide (4.3×10^{-3} mol/l), tadpoles were completely narcotized within 4 minutes. After remaining in the solution for 6 hours, and then being transferred to fresh water, they recovered partially within 5 minutes, and completely within 10 minutes.

Phthalide is soluble in 150 parts of cold water and in 45 parts of olive oil.

Coumarin, $\underbrace{C_6H_4OCOCH{=}CH}$

In solutions of 1 : 4000 (1.7×10^{-3} mol/l), complete narcosis occurs within less than 2 minutes. In 1 : 12 000 (5.7×10^{-4} mol/l), tadpoles were fully narcotized within 10 minutes. When the tadpoles were transferred to fresh water, after remaining for more than six hours in this solution, their first movements occurred after 2 minutes, and they were moving quite normally a few minutes later. When exposed to a solution of 1 : 16 000 for 30 hours or more, tadpoles continue to respond to stimuli, but move sluggishly when touched. The tadpoles recover almost immediately when transferred to fresh water.

Coumarin is soluble in 400 parts of cold water, and in 40–45 parts of olive oil.

6.5 ACETANILIDE, METHACETIN AND PHENACETIN

In contrast to antipyrine and resorcinol, these important antipyretics act as narcotics when exposed at sufficient concentration, even though they are never deliberately employed by man as narcotics. With tadpoles and other gill-breathing organisms, they can be used to produce a long-lasting narcosis without fear of fatality. It has already been mentioned that it is not possible to draw a sharp dichotomy between non-specific narcotics and antipyretics, since all non-specific narcotics at the same time lower the temperature. In addition, all of these compounds produce narcosis in plants, of the type caused by ether and unlike that caused by resorcinol.

Acetanilide, $C_6H_5NHCOCH_3$

In solutions of 1 : 1000 (7.4×10^{-3} mol/l), tadpoles and entomostraca become fully narcotized within the first few minutes. After 3–4 hours, the tadpole pulse drops to 45–50 beats per minute, but the tadpoles recover immediately upon transfer to fresh water. In 1 : 1500 (9.4×10^{-3} mol/l), complete narcosis is induced within a few minutes, and can be maintained for more than 20 hours without fatality. In 1 : 2000 (3.7×10^{-3} mol/l), tadpoles become very sluggish and move only when touched. In these solutions they survive 8–14 days or more.

Acetanilide dissolves in 200 parts of cold water and in 120 parts of olive oil.

Methacetin, $C_6H_4(OCH_3)(NHCOCH_3)(1,4)$

Tadpoles do not become fully narcotized until concentrations of 1 : 700–1 : 800 ($8.6–7.6 \times 10^{-3}$ mol/l) are reached. Narcosis can last several hours without fatality, but the circulation soon becomes rather weak. In solutions of 1 : 1000 (6×10^{-3} mol/l), tadpoles continue to respond to stimulus, but are otherwise immobile. Their circulation is only slightly weakened in the first few days, but this concentration proves fatal within about 3 days. In solutions of 1 : 2000, tadpoles only become somewhat sluggish and they can live more than eight days.

Methacetin dissolves in about 500 parts of cold water and in 250 parts of olive oil.

Phenacetin, $C_6H_4(OC_2H_5)(NHCOCH_3)(1,4)$
Solutions of 1 : 2000 (2.8×10^{-3} mol/l) produce complete narcosis in tadpoles within about 10–15 minutes. The narcosis can be maintained for more than 24 hours without fatality, and the circulation is only slightly weakened during the first 30 hours. In 1 : 3000, tadpoles move only when stimulated, and circulation remains good for 3 days, with fatalities usually occurring on the fourth or fifth day.

Phenacetin is soluble in about 1400 parts of cold water, and in 370 parts of olive oil.

6.6 ADDITIVE EFFECTS OF TWO OR MORE NON-SPECIFIC NARCOTICS

In 1864, at almost exactly the same time (the same week), Claude Bernard [1] and Nussbaum [2] observed the combined effects of chloroform and morphine, and since then, a number of experiments of such a combined narcosis have been performed. In fact, it is reasonable to partially eliminate undesirable side effects of one narcotic by the opposite side effects of another one. However, if this is impossible, then it is desirable to at least reduce the harmful side effects or render them harmless as follows: first of all, small doses of one narcotic are employed which reduce both its narcotic effect and side effects. Then, a second narcotic having different side effects is added to produce complete narcosis. Under these conditions, the various side effects of the two narcotics are acceptable in their weaker form. It seems very likely to me that the narcosis of anaesthesia of the future will depend on a practical combination of several narcotics.

According to my rather numerous experiments, the narcotic effects of two non-specific narcotics combine fairly exactly. Thus, if concentration α is required to produce complete narcosis with narcotic *A*, and concentration β is required for narcotic *B*, then any mixture of these two solutions of concentrations α and β will produce complete narcosis. For example, the quantities represented by the equation $\frac{1}{2}\alpha + \frac{1}{2}\beta$ will produce complete narcosis, although sometimes the narcotic effect of such combinations is somewhat weaker than normally expected. If one substance is insufficiently soluble for its saturated solution to produce narcosis, or if its other effects occur at much lower concentrations than the narcotic effect, then by using this additive technique, it can be demonstrated in many cases, that the substance truly acts as a narcotic. Since I have already discussed such combinations earlier in this study (p. 32), it does not seem necessary to re-examine these in detail.

REFERENCES

1. Bernard, C. (1875) *Leçons sur Les Anesthésiques et sur L'Asphyxie,* p. 226.
2. Nussbaum, Prolongation de l'anesthesié chloroformique pendant plusieurs heures. *Intelligenzbl. für. bayer. Aertze,* cited according to C. Bernard[1].

7

Inorganic anaesthetics

7.1 CARBON DIOXIDE

It has been known for some time that carbon dioxide can act as a local anaesthetic. In 1858 Ch. Ozanam [1] showed that it is also a general anaesthetic. The ability of carbon dioxide to produce general narcosis was then subsequently studied more closely by, among others, Samson [2], Paul Bert [3], and Grehant [4].

7.1.1 Carbon dioxide and natural sleep

For various reasons, the study of the narcotizing properties of carbon dioxide is particularly interesting. The phenomena of artificial narcosis have such similarities to those of natural sleep that one is forced quite involuntarily to ask the question whether or not natural sleep is caused by a narcotic-acting substance produced by the organism itself. In weighing such a hypothesis, one's thoughts turn all the more to carbon dioxide, since it would be absolutely necessary for the hypothetical compound to be able to pass out of the ganglia cells. However, this could happen only if a decrease in the concentration of the compound in the blood capillaries of the brain corresponded to a rapid change in its level in the ganglia cells; which would actually be the case with carbon dioxide, but not with a basic compound.

If I am not mistaken, it was Dubois [5] who first expressed the theory for a specific case of natural sleep, namely that the hibernation of marmots is produced by a retention of carbon dioxide in the blood. Dubois assumed that the total amount of blood is heavily saturated with carbon dioxide. In this form, the hypothesis can under no circumstances be extended to include the natural sleep of non-hibernating mammals, and even for the marmot, this hypothesis seems to me very improbable. In order to attribute a greater significance to the role of carbon dioxide in natural sleep, one must make the tentative, and still experimentally unverified assumption that, as a result of delayed cerebral circulation, carbon dioxide accumulates in the blood capillaries and the ganglia cells of the forebrain without also accumulating in the other parts of the body. According to this theory, the pressure of the carbon dioxide in the forebrain and perhaps in the central brain would also significantly increase without the average pressure of the carbon dioxide in the blood as a whole needing to increase perceptibly. This theory would mean that the state of agitation and dysponea that occurs in the event of a moderate increase in the carbon dioxide pressure in the blood as a whole, would not take place, since these undoubtedly originate in the medulla oblongata and the spinal cord. Even in this form of the hypothesis, it seems very doubtful to me that such a considerable accumulation of carbon dioxide in the brain is in any way likely, and could alone explain the phenomenon of sleep. However, it seems possible to me that such an accumulation of carbon dioxide plays a certain part in the onset and maintenance of sleep.*

The fact is that the cerebral circulation is much slower during sleep than in the waking state. Of course, it does not automatically follow from this fact that the ganglia cells and blood capillaries of the brain are more heavily laden with carbon dioxide during sleep than in the active state of the brain, since the decrease in the amount of blood that flows through the brain over a certain period of sleep is possibly compensated, or overcompensated for by an equally great or greater decrease in the production of carbon dioxide (and in the consumption of oxygen). The comparison of the ratios of the rate of circulation in the brain in the sleeping and waking states with those found in a muscle or a gland could appear favourable to this theory. However, it should not be forgotten that only a fleetingly small amount of mechanical work is associated with the activity of the ganglia cells, and that a perceptibly more vigorous metabolism of the brain during its active state does not therefore necessarily have to be assumed. If one does make this assumption, it is really because a more vigorous circulation is observed during the active state and an analogy is drawn that is of somewhat dubious reliability. There can, of course, be no doubt that in very many cases, the greater activity of the ganglia cells occurs first, and the more vigorous circulation in the brain comes next, or that the slower circulation is a result of the reduced activity of the brain. However, this fact in no way rules out the possibility that in other cases the change in circulation comes first, and the change in the intensity of the activities of the ganglia cells comes second,

* Bradbury gave a good overview of the newer hypotheses on sleep in his lecture 'On sleep, sleeplessness and hypnotics', *Lancet*, June and July 1899.

since it is general knowledge that a compression of the carotid arteries lowers the activity of the brain, or stops it completely. In former times, such a compression of the blood vessels of the neck was said to have been applied sometimes during surgical operations in order to remove pain [6]. There is reason to believe that for the most part the drawing of large amounts of blood to the skin, and away from the brain*, can be attributed to the soothing effects of warm baths and the sleep-inducing effects of wrapping up the patient in the case of certain illnesses, although the excretion of harmful compounds from the blood as a result of stimulated sweat gland activity may play a certain part in many cases.

In weighing up the various circumstances, it seems quite possible that at first the reduced activity of the brain immediately before going to sleep causes a slowing of the cerebral circulation and that this reduction in circulation rate is greater than would correspond to a possible lowering of brain metabolism during the period of reduced activity. The resulting carbon dioxide accumulation in the ganglia cells and the brain capillaries in turn reduces the activity of the ganglia cells even further until these continuously alternating reactions lead to a new state of equilibrium, sleep.

I was originally led to the above line of thought by observations of sleeplessness in the mountains. The large majority of people who come directly from lower to higher (over 1500–2000 m) elevations, sleep very poorly the first night without, however, experiencing fatigue on the next day. In completely healthy individuals, this sleeplessness quickly disappears, but for individuals with an existing tendency to insomnia, it can persist. We have learned from numerous experiments of Paul Bert that a reduction in air pressure results in a significant decrease in the carbon dioxide level of the blood.†

* It is immaterial to the analysis whether the brain actually contains less blood or not, only as long as the amount of blood flowing through the brain, or the cerebral cortex, is reduced.

† It is natural to attribute the favourable effect on the organism of the mountain climate, or of a moderately high ascent in an airship, (particularly the effect on the healthy organism, but perhaps partly also on the diseased organism) mainly to this decrease in the carbon dioxide in the blood and tissues. In fact, an increase in the alkalinity of the blood and tissues is associated with the lowering of the concentration of carbon dioxide as a result of the increase in the dissociation of carbonate. However, this increase in alkalinity in turn causes, at least judging from experiments with surviving pieces of tissue, greater activity in the cell tissues.

A short time ago, Mosso, in his book, *Der Mensch auf den Hochalpen, (Man on the High Alps),* proposed the hypothesis that altitude sickness is caused by the lowering of the carbon dioxide level in the blood. However, Mosso proceeds from the false assumption that a reduction in air pressure of less than half an atmosphere cannot cause a lack of oxygen in the blood, because the oxyhaemoglobin does not start to dissociate in air until an ambient air pressure of 238 mmHg is reached, corresponding to a partial pressure of oxygen of

Bert gives the average value for the decrease in the carbon dioxide content of the carotid arteries at a pressure of 450 mmHg, which would correspond to an altitude of 4000 m above sea level, or 14% of the normal value. It immediately becomes obvious that, if the arterial blood going to the brain is lower in carbon dioxide, the accumulation of carbon dioxide in the brain to a given concentration level will be more difficult, i.e. will demand a greater retardation of the cerebral circulation. The latter will be able to happen only through a kind of adaptation of the vascular-constricting nerves of the brain to the new conditions. It would not be surprising if this adaptation took place much more quickly in a healthy nervous system than in a somewhat disturbed one.

---

50 mmHg. Even if oxyhaemoglobin did actually undergo a noticeable dissociation at body temperature at the stated partial pressure of oxygen, which is in fact not the case (the dissociation at higher oxygen pressure is only slight), in the case of Mosso's argument, it would be a question of confusing a dynamic problem with a static one.

So far as the absorption of oxygen by the blood is concerned, it is not a matter of how much oxygen could be absorbed by the blood if a static equilibrium were established, but of how much oxygen is actually absorbed by the blood as it flows quickly through the capillaries of the lungs, since the time that it flows through is too short for static equilibrium to be established. Let us assume for the sake of argument that dissociation of the oxyhaemoglobin does not become great enough for practical consideration until the partial pressure of oxygen decreases below 70 mmHg, and let us make the further assumption that the partial pressure of oxygen in the air breathed is 150 mmHg. The partial pressure of oxygen in the alveoli would then be approximately 120 mmHg. Therefore, the difference in the oxygen pressure between the blood and the air in the alveoli until the haemoglobin essentially becomes saturated with oxygen would never fall below 50 mmHg. However, if we assume in a second case that the partial oxygen pressure of the air breathed is only 105 mmHg and the pressure in the alveoli is therefore about 75 mmHg, then the difference in oxygen pressure between the air in the alveoli and the blood until the same degree of saturation is reached in the haemoglobin as in the first case (if this could ever in fact happen) would drop to 5 mmHg, and during the entire time that the red corpuscles were being loaded with oxygen, the difference in pressure would be much lower than in the first case that we examined. The amount of this difference in pressure, however, determines the rate at which the oxygen passes from the alveoli of the lungs into the blood. Furthermore, Bert's numerous analyses of the blood gases at moderately lowered pressure provide absolute proof of a decrease in the oxygen level in the blood. According to Bert, at a pressure of 450 mmHg (90 mmHg oxygen) in the air breathed, the oxygen level in the arterial blood is lowered by about 20%. In normal individuals, altitude sickness tends to become more noticeable only at about this reduction in air pressure, which corresponds to an altitude of slightly over 4000 m above sea level, although in anaemic individuals or those with weak lungs, altitude sickness may occur, of course, much sooner. One must also not lose sight of the fact that oxygen consumption during more strenuous work is 4–8 times greater than at rest and that there are limits to even the amazing adjustment that the body makes to these greater demands by means of accelerated circulation and increased aeration of the lungs.

7.1.2 Partial pressure of carbon dioxide necessary for narcosis

After these introductory remarks, to which, by the way, I do not wish to attach any greater significance, I will proceed to explain the results of my experiments on the narcotizing effects of carbon dioxide.

Various authors have made widely differing statements on the partial pressure of carbon dioxide required in the air to produce narcosis. This is very understandable, since in fact it does not depend on this partial pressure at all, but on the concentration of carbon dioxide in the brain lipoids of the ganglia cells. This concentration is in a certain proportion to the concentration of the carbon dioxide in the intracellular fluid in the ganglia cells and in the blood plasma of the capillaries supplying them, a ratio that is governed by the partition coefficient of the free carbon dioxide between the blood plasma and the brain lipoids. If no carbon dioxide were formed within the organism itself, and if the temperature of the blood was always the same, then of course the concentration of the free carbon dioxide in the blood plasma would be in a certain proportion to the partial pressure of the carbon dioxide in the inhaled air. In fact, neither one condition nor the other is satisfied.

We will first examine warm-blooded animals, i.e. mammals and birds. Relatively vigorous carbon dioxide formation takes place in these animals so long as they are alive, so that the concentration of free carbon dioxide in the tissues and the blood rises much higher than corresponds to the partial pressure of carbon dioxide in the inhaled air. This carbon dioxide formation, however, is much more intensive in small mammals and birds than in larger mammals. For example, it is 10–15 times greater in the smaller songbirds than in humans, when equal body weights are compared. These differences are only partially balanced out by more rapid circulation and better aeration of the lungs. Another factor is the phenomenon that narcosis by carbon dioxide is accompanied by an especially large temperature decrease in the test animal. This lowering of the temperature affects the smaller animals the most quickly and intensively, and it can amount to 8, 10, or more degrees Celsius. Now it is known that the absorption coefficient of water (and also of blood plasma and tissue fluids) for carbon dioxide, as for other gases, is considerably higher at a lower temperature than it is at a higher temperature. For all these reasons, it follows that, at a given partial pressure of carbon dioxide in the inhaled air, the concentration of free carbon dioxide will be higher in the blood plasma of small mammals and birds than in the blood plasma of larger mammals and, furthermore, that the concentration of carbon dioxide during the experiment must increase at first (because of the lowering of the temperature) and that this will happen more rapidly in smaller animals than in larger ones. In any case, lowering the temperature of the animal will also reduce its metabolism and therefore also its formation of carbon dioxide. This is, in turn, partially compensated for by slower circulation and respiratory movements.

In fact, it follows from Bert's experiments that birds become narcotized, or are killed, by a lower carbon dioxide level in the air (24–28% by volume) than small mammals, and the latter become narcotized at a lower carbon dioxide level than large mammals. Of course, in Bert's experiments, the conditions were complicated by the fact that in most cases, he let the carbon dioxide content of the inhaled air increase very slowly through respiration itself, so that narcosis, or death, did not occur for hours. Only his experiments 616, 618 and 619 (*Pression Barométrique*, pp. 994–6) are free of this complication. According to these experiments, complete narcosis would occur in dogs at a carbon dioxide level of approximately 40% by volume in the inspired air. N. Gréhant [7] states that rabbits become fully narcotized within 2 minutes if they breathe an air mixture consisting of 45% by volume carbon dioxide, 35.2% nitrogen and 20.8% oxygen, and that this narcosis can last 2 hours without causing death. He states that an air mixture of this composition also produces narcosis in larger animals and humans.

7.1.3 Tadpole experiments

Most of my own experiments on the narcotic effect of carbon dioxide were again conducted on tadpoles. I used two different experimental methods. The first method was as follows: a fairly large flask was partially filled with distilled water and completely saturated by passing carbon dioxide through it for some time at a known temperature. Using a pipette that had several spherical extensions and had already been filled with carbon dioxide gas, one part by volume of the saturated carbon dioxide solution was transferred to another flask that already contained 2, 2½, 3, 3½, 4, 4½ or 5 volume parts of distilled water saturated with air or oxygen. The flask chosen was of a size that, after transferring the water saturated with carbon dioxide to it, it appeared to be filled to within 1–2 ml. The end of the pipette, which naturally must contain much more water saturated with carbon dioxide than the one part by volume that was transferred to the flask, was held well below the surface of the distilled water. The container was sealed and shaken well and then small tadpoles were placed in it. If the flask has a fairly long and narrow neck, this can all be done without any appreciable loss of carbon dioxide. Since the absorption coefficient of carbon dioxide for water is known for the temperature at which the saturation was carried out, or can easily be calculated with sufficient accuracy by interpolation, the carbon dioxide concentration of the water could be determined very easily in individual experiments.

In the other method, two or three small tadpoles, entomostraca, and other such organisms were placed in a small dish containing only enough water to permit the test animals to swim around. The dish was fastened to one end of a long knitting needle with the aid of a strip of wire netting and some thin wire and placed inside a wide-mouthed flask of about 3 litre capacity, the other end of the knitting needle was pushed through the cork of the flask. A glass tube with an exterior right-angle bend and a stopcock were also pushed through this cork. The tube was then connected to a strong aspirator and the air in the flask was diluted to the point at which the pressure dropped to 45, 50, 55 mmHg, etc., whereupon the stopcock was closed and the tube was connected to a carbon dioxide generating apparatus that had already been running for a while. The stopcock was then carefully opened and pure carbon dioxide admitted into the flask until the pressure in the flask reached the same pressure as in the room. The stopcock was then closed. As in the experiments with chloroform and ether vapours, the carbon dioxide was thoroughly mixed with the air in the flask by using the knitting needle to move the little dish up and down. This permitted the water in the dish to absorb a quantity of carbon dioxide corresponding to the known partial pressure of carbon dioxide at that room temperature. Within a few minutes, equilibrium was essentially reached between the carbon dioxide level in the water and in the air.

Experiments

At 2.41 pm on May 11, 1897, three tadpoles of *Rana temporaria* were placed in a mixture of 53 ml of water saturated with carbon dioxide at 16°C, and 51 ml of water saturated with air. The air space was 2 ml. At 2.43 pm, all three tadpoles were very sluggish, but not yet completely narcotized. At 2.45 pm, all were already entirely unresponsive to stimuli. At 2.55 pm, one tadpole was removed with the aid of a narrow glass tube and examined under the microscope. Circulation was still present, but was fairly weak. At 2.58 pm, this tadpole was transferred to fresh water. After a few minutes it was again responsive and soon thereafter completely lively. At 3.15 pm, the two remaining tadpoles were transferred to fresh water and afterwards studied immediately. The circulation was well maintained and the tadpoles recovered after a few minutes. In a second experiment, tadpoles were maintained in a solution of the same strength for 5 hours. However, they were already dead, and from their appearance, seemed to have been dead for some time. The absorption coefficient of carbon dioxide for water at 16°C is approximately 0.98. Therefore, the solution contained 0.096% by weight carbon dioxide

(0.021 mol/l) and would correspond to a partial pressure of carbon dioxide of about 400 mmHg at 16°C.

At 9.52 am on May 7, three tadpoles of *Rana temporaria* were placed in a mixture of 75 ml of water saturated with air and 25 ml of water saturated with carbon dioxide at 16°C (after correction for a small air space). At 10.00 am, the tadpoles were almost completely narcotized, and by 7.30 pm on the same day they were already dead. In a repetition of the experiment, complete narcosis occurred within 15 minutes. After 5 hours, the tadpoles were already dead. In a third experiment with a carbon dioxide solution of the same strength, two of the three tadpoles were dead after being kept for 5 hours in the solution, but the third recovered after transfer to fresh water.

Tadpoles became sluggish, but not fully narcotized, when kept in solutions formed of mixtures of 20 parts water saturated with carbon dioxide at 16°C, and 80 parts saturated with air. Most of the tadpoles tended to die, however, after 6–8 hours.

Therefore, a solution containing 0.048% by weight (0.010 mol/l) carbon dioxide can be designated as the narcotic threshold solution. It corresponds to a partial pressure of carbon dioxide of approximately 200 mmHg at 16°C.

7.1.4 Effect of temperature

A longer series of additional experiments was carried out, partially using the first method, and partially using the second. These experiments produced unmistakable proof that the partial pressure of carbon dioxide at which narcosis occurs varies with the temperature in such a fashion that, the lower the temperature, the lower the partial pressure of carbon dioxide required for narcosis. For example, the pressure at 8°C is only about 150–160 mmHg, but at 30°C, it is already almost 300 mmHg. It follows from this that the concentration of carbon dioxide in the water surrounding the tadpoles, and therefore, also the concentration of free carbon dioxide in the blood plasma of the tadpoles, that is just sufficient for narcosis varies only slightly with temperature. The concentration required for narcosis is also substantially the same for tadpoles and mammals. Thus, in the case of carbon dioxide, we find the proportions to be quite similar to those for ether and chloroform.

Paul Bert has already stated that amphibia and reptiles are 'more sensitive' to carbon dioxide, i.e. they die at a lower partial pressure of carbon dioxide in the inspired air than mammals and birds. However, probably as a result of his somewhat limited understanding of the role of the partial pressure, Bert did not discover the reasons for this behaviour. They can be found quite simply in the greater absorption capacity of the blood and cell fluids for carbon dioxide (as for other gases and vapours) at lower as compared with higher temperatures. Actually, Bert's experiments are not very conclusive, since death did not occur in his cold-blooded test animals for 1–3 days, while death occurred in the warm-blooded animals after 2–3 hours, although this occurred at a much higher carbon dioxide level in the air.

According to Theodor de Saussure [8], at 18°C, 100 parts by volume of olive oil absorb 151 parts by volume of carbon dioxide, while according to Bunsen (calculated by interpolation and using Bunsen's definition of the absorption coefficient), 100 parts of water at the same temperature dissolve almost exactly 100 parts of carbon dioxide. Thus the water/oil partition coefficient of carbon dioxide equals 1.5.

From the investigations of other authors, and from my own experiments, it has been found that carbon dioxide cannot be considered one of the better narcotics, since it causes death after a relatively short-lived narcosis. There can hardly be any doubt that it is the significant reduction in the alkalinity of the blood and cell fluids that causes the harmful effects of carbon dioxide. This reduction in alkalinity accounts for the decrease in oxygen consumption (which is readily seen in Bert's experiments, for example) and the extreme drop in temperature. On the other hand, the mechanism of the actual narcosis is probably the same as for the non-specific narcotics.

7.1.5 Mechanism of the absorption and release of carbon dioxide

Up to now I have made the tacit assumption that the passage of carbon dioxide from the cell tissues into the blood, and from the blood into the alveoli, and *vice versa*, depends entirely on diffusion processes. We know, however, that Bohr [9] proposed the hypothesis that when carbon dioxide passes from the blood into the alveoli, a special secretory (or excretory) action in the epithelia of the lungs plays a part. Similar ideas have also been expressed regarding gas exchange between the tissues and the capillaries of the circulatory system, so I cannot leave this subject entirely untouched. The following important arguments can be raised in objection to these hypotheses, at least so far as carbon dioxide is concerned: (1) this author succeeded many years ago in demonstrating by means of plasmolytic experiments that plant protoplasts are extremely permeable to carbon dioxide solutions, that this permeability is equal in both directions and is not perceptibly reduced either by cold or by narcosis of the cells. It has, however, been proven by a large number of experiments that if a compound can pass rapidly in and out through a plant protoplast, it can also pass equally quickly in and out of most animal protoplasts; (2) this author has succeeded in proving by direct experiment that carbon dioxide passes quickly in both directions through living muscle fibres; (3) Bert's experiments [10] show that if animals inhale an air mixture rich in carbon dioxide, not only does an accumulation of the carbon dioxide manufactured by the test organism itself occur, but also that during the first few minutes of the experiment, large quantities of carbon dioxide pass from the inhaled air mixture into the blood, i.e. permeate through the epithelia of the lungs in the opposite direction from normal. An experiment of Pflüger [11] demonstrates the same thing. It can also readily be seen from numerous experiments of Bert and others that the increase in the carbon dioxide level in the inhaled air corresponds to an approximately equivalent increase in the carbon dioxide pressure in the blood. This increase in pressure is, of course, not measured directly, but follows from analyses of the blood gases. As a result, the amount of the pressure increase can only be approximately estimated. It has also been found that in precisely those cases in which a really vigorous excretory action in the epithelia of the alveoli could be useful to the organism, if such action is present at all it proves to be as good as useless. At the moment, therefore, Bohr's hypothesis seems to me to be highly improbable, although not entirely refuted.

7.2 CARBON DISULPHIDE

I performed only a few experiments with this compound. Tadpoles and entomostraca become fully narcotized at a concentration of approximately 1:25 000 (5×10^{-4} mol/l), and they recover if narcosis lasts for only a short time. However, carbon disulphide cannot be regarded as one of the better narcotics, since it produces side effects that prove fatal relatively quickly. There are, however, no further data on this in my notes.

Carbon disulphide is soluble in about 1500 parts of water and is miscible with olive oil in all proportions.

7.3 NITROUS OXIDE

Few anaesthetics can claim greater interest than nitrous oxide. For one thing, the actual history of non-specific narcotics (anaesthetics) begins with this compound. In addition, it served as the starting point for one of the most important recent steps in the understanding of narcosis; i.e. the discovery that an equilibrium exists between the degree of narcosis and the partial vapour pressure of the anaesthetic in the inhaled air. Of course, at present we must add that this exists at a fixed and constant temperature. In fact, it is easy to understand how it was precisely this gaseous anaesthetic that must have guided Bert's thoughts to this subject, since he had discovered as a result of numerous experiments that the physiological effect of simple gases and gas mixtures always depends upon the partial pressures of the specific gases. Quite apart from this historical interest, however, nitrous oxide is in many respects of great significance to the theory of narcosis, partly because it is the only known narcotic that does not contain carbon, and also because it is more free of side effects than any other narcotic. In fact, it has been possible to sustain much longer lasting narcoses in mammals and humans with nitrous oxide than by means of any other anaesthetic. Narcosis in man has been successfully extended for seven hours without incident and Martin (Lyon) reported that he had maintained narcosis in a dog for 60 hours [12] without killing the animal. What more could one wish for in this respect?

Unfortunately, nitrous oxide does not produce complete and lasting narcosis in warm-blooded animals until a partial pressure of 760 mmHg is reached. Since this compound cannot take the place of oxygen, it is necessary to simultaneously inhale oxygen, preferably at a partial pressure of about 150 mmHg. In practice, it has been found that even higher partial pressures of nitrous oxide and oxygen are advantageous. For example, a gas mixture is inhaled consisting of 15% by volume of oxygen and 85% by volume of nitrous oxide at a total pressure of about 950–980 mmHg.

This naturally requires the use of a complicated apparatus, all the more so in view of the fact that the operator himself has to remain in compressed air during the operation. These circumstances have restricted the general use of nitrous oxide as an anaesthetic up to now, in spite of its great advantages. Perhaps combining nitrous oxide with another non-volatile or less volatile narcotic (possibly a urethane) will eventually prove expedient.

R. Dubois [13] and many French physiologists theorize that the mechanism of action of nitrous oxide differs entirely from that of the organic anaesthetics; Dubois calls it the only representative of the 'special functional anaesthetics', as opposed to all other anaesthetics ('general anaesthetics'). To support his theory, Dubois advances the idea that plants are not narcotized by nitrous oxide. This statement is quite true, if plants are exposed to a nitrous oxide partial pressure of only one atmosphere. This fact, which could easily be predicted, however proves absolutely nothing since a 6–10 times higher concentration of ethyl ether, chloroform and most non-specific narcotics is required to produce narcosis in plants relative to narcosis in higher animals. If we also take into account the fact that at 15°C, and at the same partial pressure, about twice as much nitrous oxide is absorbed by water and aqueous solutions than at 38°C, then narcosis would not be expected to occur in plant cells, or in protozoa, ciliary cells, etc., until a nitrous oxide partial pressure of 3–5 atmospheres is reached. Due to the lack of suitable apparatus I have unfortunately not been able to conduct any experiments under extreme pressure. Nevertheless, I do not doubt in the least that narcosis in plant cells could be successfully achieved with nitrous oxide under higher pressure. Living plant cells, red blood corpuscles, and other cells are very permeable to solutions of nitrous oxide, as has been demonstrated by plasmolytic methods.

By analogy, it would be anticipated that tadpoles would be completely narcotized at a temperature of 15°C at a nitrous oxide partial pressure of about 400–500 mmHg. They appear to become narcotized, however, only at somewhat higher pressures. In the notes of my experiments on this subject there is, however, the remark that the original solution was probably not completely saturated with nitrous oxide. I hope to repeat the experiments using the second of the methods mentioned under carbon dioxide.

According to Bunsen, the absorption coefficient of nitrous oxide for water at 20°C is 0.67. According to an earlier statement of de Saussure [14], 100 parts by volume of olive oil absorbs 150 parts by volume of nitrous oxide at 18°C. The oil/water partition coefficient is therefore about 2.2.

In considering the data provided above on nitrous oxide, it appears to me beyond a doubt that the mechanism of this compound, as far as its primary narcotic action is concerned, agrees entirely with that of the organic non-specific narcotics, but is particularly free of side effects since it remains chemically entirely unchanged. This is not to say that its presence in the cell fluids leaves the rate of the normal chemical processes in the organism entirely unchanged. The fact that the narcotic effect of nitrous oxide ceases almost instantly when it is no longer being inhaled can easily be explained as follows: the pressure of this gas in the blood at the concentration required for narcosis is one whole atmosphere, and also its solubility is rather high. Therefore, the absolute difference in concentration between the portions of the solution lying closest to the respiratory epithelium and the portions that are furthest removed from it will be quite considerable (the whole distance amounts to only a few micrometres), until the concentration has fallen far below the level required to maintain narcosis. Since the partition coefficient (brain lipoids/water) is probably only on the order of 2, then the nitrous oxide passes out of the ganglia cells into the brain capillaries as soon as the nitrous oxide concentration falls in the latter.

Like this author, H. Meyer also considers the mechanism of nitrous oxide induced narcosis to be the same as that induced by non-specific organic narcotics.

REFERENCES

1. Ozanam, C. (1858) *Acad. des Sc.*, Feb. 25, Cited acc. to Dastre (1890) *Les Anesthésiques*.
2. Sanson, A.E. (1865) *Chloroform, its Action and Administration*, John Churchill, London; (1864) On the Action of Anesthetics and on the Administration of Chloroform, *Medical Times and Gazette*. Cited according to Bernard, C. (1875) *Leçons sur les Anesthésiques et sur l'Asphyxie*.
3. Bert, P. (1878) *La Pression Barométrique*, G. Masson, Paris.
4. Gréhant, N. (1887) Anesthésie des rongeurs par l'acide carbonique; application du procedé de Paul Bert, *C.R. Soc. Biol. Mém*, Paris, **39**, pp. 52–4, 153–4.
5. Dubois, R. (1896) Étude sur le mecanisme de la thermogenese et du sommeil chez les mammiferes: *Physiologie Comparée de la Marmotte*, Masson et cie, Paris, p. 253. Cited according to Mosso, *Der Mensch auf den Hochalpen*.
6. Cited according to Bernard, C. (1875) *Leçons sur les Anesthésiques etc.*, Librairie J.-B, Baillière et Fils, Paris, p. 35.
7. Gréhant, N. (1887) *Soc. Biol.*, Jan. 29 and Mar. 12. Cited according to Dastre, *Les Anesthétiques*.
8. de Saussure, T. (1814) *Über die Absorption der Gasarten*, etc., *Gilb. Ann.*, **47**, 169.
9. Bohr, (1891) *Skandin. Arch. Physiol.* (Leipzig), **2**, p. 236.
10. Bert, P. (1878) *La Pression Barométrique*, G. Masson, Paris, pp. 994–5, especially experiment 618.
11. Pflüger, E. (1868). Ueber die Ursache der Athembewegungen etc., *Pflg. Arch. Physiol.*, **1**, pp. 61–106.
12. Cited according to Dubois, R. (1894) *Anesthésie Physiologique et ses Applications*, George Carrbe, Paris, p. 121.
13. Dubois, R. (1894) *Anesthésie Physiologique et ses Applications*, George Carrbe, Paris, p. 116.
14. de Saussure, T. (1814) Beobachtungen über die Absorption der Gasarten durch verschiedene Körper. *Gilb. Ann.*, **47**, 169. (Translation of one of the lectures of the *Naturf. Ges. in Gent*, April 16, 1812).

8

Action of basic narcotics and basic compounds

It has already been emphasized in the general section of this study that although the basic narcotics usually act in a rather different fashion from the non-specific narcotics, nevertheless, these two groups overlap in various ways. It therefore seems appropriate to touch on this subject briefly [1].

8.1 CLASSIFICATION OF THE BASIC ORGANIC COMPOUNDS ACCORDING TO THEIR DEGREE OF ALKALINITY

Disregarding the acid amides and amido acids, the second of which in any case cannot penetrate most living cells at all, or can penetrate only as a result of a particular activity of the protoplasm, then the organic compounds of rather distinct basic character can be classified within three groups according to their degree of basicity. Within the first group belong the quaternary amines, NR_4OH, where R_4 refers to four equivalent or non-equivalent substituents such as CH_3, C_2H_5, etc. These quaternary amines are all very strong bases of similar basicity to potassium hydroxide and sodium hydroxide,, i.e. in fairly dilute aqueous solution they are already rather completely dissociated electrolytes. They are for the most part much less readily soluble in ether and similar solvents than in water, and their partition coefficient between ether and water very much favours water (ions are almost entirely insoluble in ether, the actual partitioning corresponding essentially only to those molecules that are not dissociated electrolytes). Accordingly they cannot permeate most cells, so long as these are still undamaged, but they seem to be absorbed by the ends of the motor nerves, since most of them exert an effect like that of curare. However, basic aniline dyes of the structure of quaternary amines, e.g. methylene blue, readily penetrate living cells. We will not be concerned with this group any further.

A second group of organic bases have a basicity in rather dilute aqueous solution similar to ammonia, i.e. the individual members of this group are electrolytes which are dissociated in such aqueous solutions from only tenths of a percent to several percent. This group consists of aliphatic primary, secondary and tertiary amines including cyclic amines in which the nitrogen atom is in a partially or entirely hydrogenated ring, and also, naturally, those amines in which the nitrogen atom is bound to a ring system only by means of a hydrocarbon chain. Almost all of these compounds, so long as they are monoamines, are rather soluble in ether and similar solvents. Those with higher molecular weight are usually more soluble in ether than in water. In the free state, but not as their salts, they readily penetrate living plant and animal cells, so long as these cells remain undamaged. These amines do not have a caustic effect, which depends upon the concentration of the hydroxyl ions, until they reach much higher concentrations than the quaternary amines. This group includes most of the physiologically more important alkaloids.

The third group includes the true aromatic amines and the cyclic amines in which the amine nitrogen forms part of an unsaturated ring, i.e. on the one hand, compounds such as aniline and diphenylamine, and on the other hand, compounds such as pyridine, quinoline, isoquinoline, and papaverine. All of these compounds are very weak bases, whose salts, even in moderately diluted aqueous solution with even the strongest acids, are very clearly hydrolysed. The aqueous solutions of these salts produce a distinctly acid reaction, for example, with litmus. The free bases are rather readily soluble in ethyl ether, higher monohydric alcohols, etc., and thus very rapidly penetrate living plant and animal cells. It is members of this group that bridge the gap between the non-specific narcotics, and the narcotics of more distinctly basic character. In the case of some alkaloids containing two nitrogen atoms, one of the nitrogen atoms belongs to a saturated ring, and the other to an unsaturated ring. In these cases, the basicity of the one basic group is considerable, and that of the other group is weak. The salts of these alkaloids with one acid equivalent are accordingly barely perceptible in aqueous solution, but the salts with two equivalents, on the other hand, are subject to fairly strong hydrolytic dissociation, causing only the second acid equivalent, however, to be noticeably dissociated. Examples of such alkaloids are, among others, nicotine and the cinchona alkaloids. From the physiological point of view, these alkaloids would be classified within the second group.

8.2 FORMATION OF SALTS WITH CELL PROTEINS

If water soluble alkaloid salts are introduced into the organism, the salts are almost completely decomposed by the alkali of the blood, lymph, etc., so that these salts exist in fact as the free alkaloids. Since the free alkaloids are usually more soluble in the brain lipoids than in aqueous solution, they will accumulate up to a certain point within the brain lipoids, just as in the case of the non-specific narcotics. Thanks to their basic properties, however, the alkaloids, or in more general terms, the amines, are also capable of forming salt-like compounds with the various proteins in cells. In general, an alkaloid, or other amine will be bound more rapidly forming salts with proteins, etc., in proportion to the strength of its basicity, since the hydrolysis of its salts decreases with increasing basicity of the amine base, assuming equal solubility of these salts. It is not just a question, however, of the basicity of the corresponding bases and the acidity or binding strength of the various proteins, but equally important, is the solubility of the resulting alkaloid or amine protein salts. If, for example, a basic compound of the third group binds to a cell protein belonging to a certain type of tissue to form a product, which despite its low basicity, is not readily soluble in the plasma fluid of the corresponding cells, then, the major part of this basic compound may be converted into the bound state, since the hydrolytic dissociation affects for the most part, only that fraction of the compound present in solution. For all basic compounds capable of readily penetrating living cells, there appears to exist an *a priori* possibility, that, as a result of their capacity to form salt-related compounds with any cellular protein, they exert a physiological effect at concentrations that are much too low to act like non-specific narcotics. However, the probability that the first effect mentioned will occur before the second is less likely the weaker the basicity of the corresponding compound, and the greater its (oil/water) partition coefficient. These conclusions have been fully confirmed by experience, as the following data will show.

8.3 ACTION OF SOME VERY WEAK ORGANIC BASES

Diphenylamine

This very weak base, whose salts in aqueous solution become far more dissociated than aniline salts, acts almost exactly like a non-specific narcotic. In solutions of 1:600 000 (3.7×10^{-5} mol/l) 13 mm tadpoles became completely narcotized within 1 hour without any particular side effects. After 6 hours, the circulation was still good, but this solution proved fatal after 20 hours. The concentration was probably higher than was absolutely necessary for narcosis. Diphenylamine dissolves in about 10 000 parts of water and is very soluble in ether.

Aniline

I would like to provide here in note form the records of some experiments on the effects of this compound.

Experiment one: At 8.51 pm on March 28, 1896, several tadpoles of *Rana temporaria*, some with exterior gills, some with interior gills, were placed in a 1 : 500 (0.022 mol/l) solution of aniline. They were very lively during the first minute; at 8.53 pm they were all completely paralysed and unresponsive to stimulus. Transferred them to fresh water at 9.00 pm; they swam around after only 2 minutes just as vigorously as before, as though nothing at all had happened. At 9.09 pm placed them in 1 : 2000 (0.0054 mol/l) aniline. They were still lively at 10.00 pm, but movements not quite normal; the tadpoles were quite agitated and had a tendency to mild spasms. The same on the following day. At 9.15 pm on March 31, i.e. after almost 72 hours, they still moved around, but sluggishly. The experiment was discontinued.

Experiment two: At 10.45 am on April 23, four tadpoles of *Rana temporaria* were placed in 1 : 1000 aniline. They were rather strongly agitated for the first three minutes; at 11.10 am they were still moving, the movements sluggish but fairly normal in character. At 3.25 pm two tadpoles were completely paralysed, the other two were very sluggish. At 8.25 pm two tadpoles were still moving when stimulated. At 8.15 am on April 24, two tadpoles were dead, a third unresponsive to stimulus but alive and the fourth still fairly easily stimulated. At 8.35 pm on April 24, all were dead.

Further experiments have shown that tadpoles become narcotized in solutions of 1 : 800 within several hours without any ill-effects. While exposed to the aniline solution, the blood corpuscles in the circulating blood undergo a rather strong deformation. This effect, however, is not due to the formation of a pigment, which according to reports in the literature, occurs in mammals after exposure to aniline. Aniline is soluble in about 30 parts of water and is miscible with olive oil.

Dimethylaniline

This compound has a purely narcotic effect on tadpoles. In 1 : 6000 (1.4×10^{-3} mol/l) tadpoles and entomostraca become fully narcotized within 1–2 minutes. After staying in the solution for more than an hour, they recover in a few minutes when transferred to fresh water. In solutions of 1 : 10 000 (2×10^{-5} mol/l) tadpoles do not respond to stimulus after 3–5 minutes; they can be left for more than 20 hours in this solution without fatality. In solutions of 1 : 12 000, tadpoles became immobile, but respond to stimuli. After 48 hours in solutions of this concentration, the tadpoles recovered within 3 minutes upon transfer to fresh water. Dimethylaniline is soluble in about 800 parts of water and is miscible with oil in all proportions.

Pyridine

This compound also acts substantially like a non-specific narcotic. In solutions of 1 : 500 (0.025 mol/l) tadpoles became immobile within 1 minute, but remained responsive to stimuli for a few more minutes. After 1 hour and 30 minutes, the tadpoles in one experiment did not respond to stimulus, their circulation becoming weak, and their pulse dropping to 40 beats per minute, with only a small amount of blood being circulated with each heartbeat. After another 3 hours, no further circulation could be detected. When transferred after that to fresh water, two of the three tadpoles moved rather well when stimulated, and 1 hour later, these two swam around vigorously, but the third tadpole was dead. On the following day, the two tadpoles were still just as lively and they had eaten up most of the third dead tadpole. In solutions of 1 : 1000 (0.013 mol/l) tadpoles no longer moved spontaneously after 5 minutes, but remained easy to stimulate for 5–6 days, with the circulation suffering only slightly in this solution. Even after being kept in this solution for several days, the tadpoles recovered in a few minutes as soon as they were placed in fresh water. No deformation of the red corpuscles was observed in the circulating blood of tadpoles that were kept in pyridine solutions.

Pyridine mixes with both water and olive oil. The oil/water partition coefficient equals 0.4.

Quinoline

In a solution of 1 : 4000 (1.9×10^{-3} mol/l) all spontaneous activity ceased within 2 minutes. Within about 5 minutes, the tadpoles no longer responded to stimulus, but their responsiveness returned again after being placed in fresh water for 2–3 minutes. In solutions of 1 : 8000 (8.7×10^{-4} mol/l), sensitivity was lost within 10–15 minutes. After 3–4 hours the circulation was yet good. After 8 hours, on the other hand, no circulation could be detected. After transfer to fresh water, the sensitivity of one tadpole returned in about a half hour, and on the following day, two or three of the tested tadpoles were moving completely normally. In solutions of 1 : 8000*, tadpoles continued to respond somewhat to stimuli for 2–3 hours. Later, they became completely unresponsive,

* Editor's note: this concentration appears to be inconsistent and may be an error in the original edition. A second dilution to 1 : 16 000 makes more sense with respect to the findings.

though the circulation remained excellent for 8–10 hours; nevertheless, the tadpoles usually died within 20 hours.

Quinoline is soluble in about 200 parts of water and is miscible with oil.

Antipyrine $N(CH_3)N(C_6H_5)COCH{=}C(CH_3)$

Antipyrine occupies a somewhat peculiar position. It is derived from the isopyrazolone ring, and can be compared to pyridine and other compounds to the extent that it is only a very weak base whose salts undergo strong hydrolytic dissociation in aqueous solution. Although the isopyrazolone ring contains only one double bond, this ring cannot be considered partially saturated, since the one double bond is the only possible one. The weak basicity of this compound is therefore not an exception to the general rule.* In contrast to acetanilide, antipyrine is not a true narcotic. Both compounds are known human antipyretics, but their mechanism of action is certainly not the same. In ¼% (1.13 mol/l) antipyrine tadpoles remained very mobile for 24 hours, with a slight tendency to weak quivering spasms. After 36 hours, the tadpoles became sluggish and their movements uncoordinated, with death usually occurring on the second or third day. Antipyrine dissolves in less than its own weight of cold water, in about one part alcohol, in 50 parts ether, and in 55 parts of olive oil.

From the data that I have given, it can be seen that all of these weak bases are very similar to the non-specific narcotics in their mechanism of action, and only antipyrine differs somewhat.

8.4 ACTION OF SOME STRONGER ORGANIC BASES

I will now proceed to the stronger, alkaline reacting bases, although I will mention only a small selection of these. All of the data refer to the free bases.

Coniine

This compound is known to act primarily on the ends of the motor nerves, but in the lower animals, it also acts simultaneously on the central nervous system. From a purely external point of view, its initial effect is almost indistinguishable from that of a narcotic. Tadpoles are unsuitable organisms for detailed analysis of the effects of coniine. Nevertheless, experiments with tadpoles are instructive in another respect, i.e. determining the effect of concentration.

* Editor's note: for information about the influence of double bonds on the basicity of heterocyclic compounds, see Perrin, D.D., Dempsey, B. and Serjeant, E.P. (1981) pKa *Prediction for Organic Bases*, Chapman and Hall, London.

In solutions of 1 : 2000 (1.0039 mol/l), tadpoles became completely or almost completely paralysed within 2 minutes, and at the same time, the epithelia of the body were strongly affected by the caustic action of coniine. Within 10 minutes, the whole body was becoming decomposed as a result of this caustic action. In solutions of 1 : 6000, severe paralysis occurred within 3–4 minutes, interrupted by occasional attacks of spasms. After 10 minutes, paralysis was complete and the epithelia of the tail were strongly affected by the caustic action of the solution. In solutions of 1 : 12 000 of coniine, the epithelia were gradually, but rather slowly decomposed and the tadpoles remained somewhat responsive to stimulus for about 20 minutes. In solutions of 1 : 24 000 (3.3×10^{-4} mol/l), responses stopped within 30 minutes, but the circulation remained rather normal for longer, and the epithelia were not obviously affected for several hours. Death followed within 6–15 hours. If the tadpoles were transferred to fresh water while their circulation was still good, most recovered completely, although much more slowly than after narcosis induced by a non-specific narcotic whose effects were experienced at the same concentration. When the solutions were further diluted, the action of coniine quickly decreased. At 1 : 40 000, tadpoles can remain alive for several days and show only slight or barely perceptible paralysis. In solutions of 1 : 80 000 and 1 : 100 000, tadpoles can survive for weeks without showing the slightest signs of disturbance.

Coniine dissolves in 90 parts of cold water (it is less soluble in hot water) and is miscible with olive oil in all proportions. The partition coefficient favours oil so much that in solutions of 1 : 25 000, coniine could already produce a narcosis like that produced by the non-specific narcotics, if the differing sequence of symptoms is disregarded. In any case, it is the capacity of coniine to form salts with proteins that is responsible for its major action. This is even more obvious with the next compound to be discussed.

Nicotine

The experimental results with this compound are so instructive that I will reproduce a rather complete account of one experiment.

Experiment At 9.45 am on May 4, 1895, three tadpoles of *Rana temporaria* were placed in 1 : 5000 (1.23×10^{-3} mol/l) of nicotine. At 9.56 am, two tadpoles no longer responded to stimulus; the third barely responded and reacted with only a faint jerk. One minute later, all three were completely unresponsive. At 10.17 am there was still no noticeable caustic effect from the solution. At 10.21 am, the circulation was still excellent. At 11.07 am, corresponding to 1 hour and 13 minutes from the start of the experiment, the circulation was still good, and the epithelia were not affected. At 11.08 am, two of the tadpoles were transferred to fresh water. At 11.15 am, they still did not move or respond to stimulus, nor at 11.35 am. At 12.05 pm, they still did not respond but the circulation was still very good. At 12.18 pm, one of the tadpoles became faintly responsive, and soon after this made spontaneous swinging movements with its tail. At 2.25 pm this tadpole had almost completely recovered. The remaining one was easy to stimulate, but did not move much. The tadpole left in the nicotine solution was now dead, but the body epithelia were not affected. At 4.00 pm, one of the tadpoles transferred to fresh water was entirely vigorous and healthy, and subsequently behaved quite normally as well. The remaining tadpole which had been transferred to fresh water worsened after 2.25 pm in that it could move less and its circulation became weaker; it finally died after a while.

In solutions of 1 : 20 000 (3.1×10^{-4} mol/l) of nicotine, all spontaneous movements ceased after 5–10 minutes, with responsiveness disappearing in 15–20 minutes. The solution was fatal within less than 8 hours. In a 1 : 50 000 solution, tadpoles moved spontaneously for more than an hour, although their movements were of very short duration and the tadpoles appeared to tire very quickly. In one experiment, a tadpole maintained its responsiveness for more than 4 hours, and when transferred to fresh water after 10 hours, recovered completely in 24 hours, even though it was unresponsive within the first few hours. Two other tadpoles in the same experiment lost their responsiveness within 3 hours and did not recover. In solutions of 1 : 80 000 nicotine, tadpoles behaved almost normally for days, although their movements seemed to drag somewhat. In 1 : 100 000 (6.2×10^{-5} mol/l) tadpoles can survive for 10 days or longer without showing any noticeable signs of disturbance.

In the case of both nicotine and coniine, their effects vary somewhat depending upon the type of amphibia and their age, i.e. the concentrations required to produce certain effects are not always identical for different species and different ages of tadpoles. In the case of some other alkaloids, there are yet much greater differences in this respect.

Nicotine is miscible in all proportions with water and is soluble in about 4 parts of oil, with the partition coefficient favouring water. The possibility of a mechanism of action like that of the non-specific narcotics is therefore entirely excluded in the case of nicotine. According to the known experiments performed with adult frogs, nicotine appears to act on the motor nerve endings, directly on the muscle fibres, and on the central nervous system.

Sparteine

In solutions of 1 : 4000 this compound exerts a highly caustic effect on the epithelia and all the other tissues of the tadpole. The protoplasm of the cells is largely decomposed within a few minutes. In 1 : 12 000, this caustic action is very weak and tadpoles can survive in these solutions for some 20 hours. The tadpoles do not exhibit any unusual behaviour, except for a little agitation. In 1 : 24 000, tadpoles behave quite normally for weeks.

Morphine

This compound penetrates living plant and animal cells much more slowly than all the other better known alkaloids (I do not include among the alkaloids ecgonine, arecaidine and similar compounds that are actually amino acids and do not perceptibly penetrate the living cell, and therefore, have no effect). Nevertheless, in comparison with many other compounds, the penetration of morphine can be considered moderately rapid. A saturated solution of morphine (1 : 2000 or 1.65×10^{-3} mol/l) produces almost no effect on tadpoles; at most, after some time, a certain stimulation is noticeable. This saturated solution does not possess a caustic action. Therefore, morphine, although its nitrogen is contained within a saturated ring, must still belong to the weaker bases of this group.

Thebaine

(Dimethyl morphine)

This produces complete narcosis in tadpoles (the records of my experiments with this compound have been lost).

Strychnine

I would like to touch briefly here on this compound because it possesses a particularly high degree of certain peculiarities of the mechanism of action of the alkaloids (or of the distinctly basic compounds in general). For example, although strychnine in solutions of 1 : 50 000 and 1 : 80 000 after a while causes strong clonic spasms in older tadpoles, even in solutions of 1 : 20 000 it has absolutely no effect on tadpoles that have just hatched out of their eggs. The striation of the voluntary muscle fibres is, of course, not yet formed in such young tadpoles, but the tadpoles can nevertheless contract. If the tadpoles are left in the strychnine solution, they grow at first like normal tadpoles. When they have reached a length of about 9 mm, they behave exactly as if they had been intoxicated with curare. Thus, they are completely paralysed, without ever having exhibited spasms, while their heartbeat is almost normal. The tadpoles can survive for several days in this condition and they reach a length of 12–13 mm. If the tadpoles are partially intoxicated with strychnine and then transferred to fresh water, it takes an extraordinarily long time before they recover completely, although their circulation remains strong and strychnine can pass very rapidly in both directions through living plant and animal cells.

8.5 SIMILARITIES AND DIFFERENCES IN THE ACTION OF NON-SPECIFIC NARCOTICS AND ORGANIC BASES

These examples will suffice to illustrate some of the more obvious similarities and differences in the phenomena of intoxication by non-specific narcotics, on the one hand, and by the stronger, basic-reacting organic compounds on the other hand. In the case of both groups, we are referring to substances that are rather soluble in cholesterol–lecithin mixtures and that accordingly pass easily in and out of living protoplasts. Furthermore, we are dealing with effects that (if they have not gone too far) can be reversed by lowering the concentration of the relevant compound in the blood, or in the solution in which the animals are maintained. However, although with the non-specific narcotics the removal of the narcotic from the solution usually produces a very rapid return to normal life signs (as long as the circulation has not been weakened too much), in general, detoxification requires a much longer time with the basic compounds. Of course, there are great differences between the individual basic compounds in this respect, so that even here there is overlapping of the two groups of toxins. A further difference in the behaviour of the non-specific narcotics and the stronger basic narcotics is that in the case of the former, the concentration of a certain compound in the blood plasma, or in the intracellular fluid of the affected cells, that is required to produce a certain effect is approximately the same in closely related animal groups. Even in groups as far apart as amphibia, entomostraca and mammals, the concentration required for narcosis is often almost the same. If we disregard fungi, even among different plants the concentration of a given non-specific narcotic usually does not vary by more than about 20% from an average level. The situation is quite different for the compounds of marked basic character. In this case, the concentrations that are required to produce a certain effect are extraordinarily divergent in closely related organisms. For example, the forms of spirogyra that belong to the type *Spirogyra orthospira* are killed by much lower concentrations of almost all alkaloids than the forms that are closely related to the *Spirogyra communis*, and all spirogyra species are more sensitive to alkaloids than the majority of other algae. The lethal concentration of strychnine and many other alkaloids is more than 50 times lower for *Spirogyra orthospira* than for the root hairs of *Hydrocharis morsus ranae* and for the cells of numerous other potted plants, although the corresponding alkaloids penetrate all these cells with about equal ability. In the case of tadpoles and adult amphibia, morphine has only a very slight effect, even in large doses, whereas in humans the maximum doses (*dosis maxima simplex* and *dosis maxima pro die*) of morphine hydrochloride are considered to be 0.03 and 0.1 g, respectively. A 0.1 g amount of free morphine distributed throughout the body fluids of an adult would correspond to a concentration of only 1 : 400 000, even disregarding the fact that it is partially absorbed into the fats, etc. Although morphine may represent an extreme case, in almost every case, the concentrations of alkaloids in the blood plasma that are required to produce a certain complex of symptoms are much higher in amphibians than in mammals, and even among mammals there are great differences in these concentrations.

8.6 COMPLEXES OF ORGANIC BASES WITH TANNINS AND PROTEINS

There is a ready explanation for the first difference mentioned between intoxication by non-specific narcotics and by basic compounds. Thus, under conditions in which the concentration of an external toxicant is decreased, the equilibrium changes much more slowly for basic compounds with respect to this state. In fact, when such basic foreign compounds are removed from the partially affected cells, this is caused not only by exosmosis, but also by processes related to chemical dissociation and solution. For example, if the concentration of a free alkaloid in the blood plasma and the intercellular lymph decreases, then at first a fraction of the alkaloid present in the free state is lost by exosmosis from the cell tissues. This decrease in concentration of the free alkaloid within the cells results in further dissociation of the protein–alkaloid complexes. It is not possible, at present, to determine whether this dissociation is only for those protein–alkaloid complexes in solution, or at the same time for those complexed molecules that are within the undissolved fraction which is in a somewhat swelled state. This uncertainty reflects our lack of knowledge about whether such a swelled substance behaves more like one that is in a bound non-swelled state, or like one in solution. The most reasonable assumption is that in this regard, such a swelled substance behaves as if it is a bound non-swelled substance with a large surface area, i.e. like a complexed material whose surface area increases in proportion to the degree of swelling. In all cases, the dissociation of the protein–alkaloid complexes will take place more rapidly along with a more rapid exosmosis of the free alkaloid from the affected cells as the protein–alkaloid complexes are more soluble and more highly swelled. Much can be learned about these relationships from the study of the response of tannin-rich plant cells when first placed in a very dilute, non-fatal solution of an alkaloid or other basic compound, and then transferred to a more dilute solution of the test compound or to fresh water once equilibrium has been reached. The tannin-complex precipitate first formed within the cell fluids is always slowly dissolved upon transfer of the cells to fresh water. As a result, the complete removal of the basic foreign compound from the cells always requires more time than that for a non-specific narcotic. Even in these cases, the time required varies widely depending upon the specific alkaloid. Thus, in the case of the rather soluble caffeine–tannin complex, cells, under favourable conditions, are free of caffeine within less than an hour after transfer to fresh water. However, the removal of strychnine from cells can require an extraordinarily long time, i.e. days or even weeks, since the tannin–strychnine complexes are extremely insoluble. Since living cells are very readily permeated by both caffeine and strychnine, it can actually be a function of only the variable degree of dissociation of the tannin complexes and the very different solubilities of these complexes.

Complexes of alkaloids with tannins and proteins are even more similar since it is not only a matter of partially dissociated complexes, but also of complexes that are amorphous in the complexed state, and somewhat swelled in the presence of water.

I would also furthermore like to mention that those complexes of cell proteins with very weak bases such as caffeine, can dissociate very quickly when the base is removed from the blood. For example, if small tadpoles are placed in 0.2–0.25% solutions of caffeine, their musculature becomes quite rigid within a few minutes. If these tadpoles are transferred to fresh water after being maintained for a short time in these solutions, the rigidity of the muscles disappears within a very short time, and the tadpoles recover completely. There seems to me to be no doubt that the stiffening of the muscles in the solution results from a combination of the caffeine with one of the proteins of the muscle fibres. In solutions of caffeine of less than 1 : 800, only the nervous system of tadpoles is affected to any extent.

8.7 RELATIONSHIP BETWEEN THE CONSTITUTION OF AN ORGANIC BASE AND ITS PHYSIOLOGICAL EFFECT

It is not easy to explain the degree to which the intensity of the effects of basic compounds varies among closely related organisms, or at least for organisms of the same type. In some cases, this behaviour results from the compound in question being chemically transformed prior to reaching the circulatory system. Thus, for example, ammonia is very rapidly converted to urea in the liver of mammals, and as a result, spasms are seldom produced in mammals, although these are produced immediately in tadpoles by very dilute solutions of ammonia. In many cases, however, this explanation is not possible, and the assumption must be made that in different species of animals, the cell proteins that correspond histologically and physiologically to each other are somewhat different. Extremely small differences in the chemical structure of the corresponding proteins might produce rather extensive differences in the solubility and in the swelled condition of their alkaloid complexes. I am completely inclined to the opinion that the pharmacological effect of basic compounds depends only to a slight degree on their chemical constitution as such, and that it is primarily a question of certain physical properties such as solubility and swelling capacity of the compounds of the base relative to the cell proteins. For example, two alkaloids of chemically different constitution will exert the same pharmacological effect if they form compounds within the same tissue organelles and with the same cell constituents (cell proteins) of about equal solubility and swelling capacity. If, in some cases, a closer relationship seems to exist between the constitution of a group of alkaloids and their pharmacological effect, this could result only from the fact that the relevant alkaloids form complexes with the same cell proteins, and that these complexes have about the same physical properties. In general, however, the solubility of a salt-like complex is related in only a very loose degree to its chemical constitution, so that it cannot normally be predicted quantitatively. This is not the place, however, to delve further into this subject. I only wished to explore in greater detail differences in the mechanism of action between non-specific narcotics and basic compounds.

REFERENCES

1. See also this author's work (1896) 'Uber die osmotischen Eigenschaften der Zelle in ihrer Bedeutung für die Toxikologie und Pharmakologie. *Vierteljahrsschr. Naturforsch. Ges. Zürich,* **41**. Reprinted without change in (1897) *Zeitschr. Physik. Chem.*, **22**, 189–209.

9

Conclusion

Upon examination of the numerous data reported in this study on narcosis, as well as the facts already assembled by Hans Meyer, there can hardly be any doubt that the action of non-specific narcotics depends upon the fact that these compounds dissolve in the brain lipoids (or the plasma lipoids). This property generally leads to the accumulation of both the moderately strong as well as stronger narcotics at this site, with changes in the physical state of the brain lipoids induced by their presence. The brain lipoids can no longer completely fulfil their functions in the nerve cell – I am referring here only to the effect of narcotics on the nervous system – or they somehow disturb the functions of other components of the neurons. These studies also demonstrate with certainty that the strength of narcotics depends primarily on their relative solubility in water and the brain lipoids, although it is too soon to say that all non-specific narcotics produce the same degree of narcosis if they have accumulated in the same molar concentrations in the brain lipoids. It is rather unlikely that such a statement is entirely valid, since the physical state of the brain lipoids is not altered to the same degree by the absorption of equimolar quantities of different compounds. The quantity of foreign molecules absorbed, however, may not be without significance. In any case, the above statement probably corresponds approximately to the actual conditions.

Recognition of this fact is certainly an important step in the development of a comprehensive theory of narcosis. This by no means, however, affords the same degree of satisfaction, for example, as the discovery that in carbon monoxide poisoning, at the same partial pressure of carbon monoxide and oxygen, i.e. the same osmotic pressures of the corresponding free gases, carbon monoxide–haemoglobin represents a far less dissociated compound than oxyhaemoglobin. On the other hand, perhaps the former knowledge will one day be employed to provide a much deeper insight into the workings of the cell than that gained by the explanation of carbon monoxide poisoning. Thus, although haemoglobin plays an extraordinarily important role in the functioning of all vertebrate organisms, it appears to have the same kind of relationship to erythrocytes as that of the cell fluid to the protoplast of the plant cell. More precisely, it appears to be only a compound dissolved within the intracellular fluid of erythrocytes, as opposed to being a constituent of the organized protoplasm of the erythrocyte framework. In contrast, the brain lipoids, which perhaps should be designated the plasma lipoids, serve as an integral part of the protoplasm of all plant and animal cells, and are second only to the cell proteins in their importance to the life of the cells. It is not unlikely that the brain lipoids have a role in defining the physical properties of the protoplasm that is equal to or even greater than that of the cell proteins themselves. The modification in the physical state of brain lipoids in response to the absorption of foreign compounds, occurs in addition, regardless of how this change affects the normal workings of the cell. This represents a common starting point for the primary effects of the majority of organic compounds, which act primarily as non-specific narcotics.

One of the most important functions of the brain or plasma lipoids is control of general osmotic properties, which are common to most plant and animal cells. Thus, the question next arises whether a noticeable disturbance of these osmotic properties takes place during narcosis. This question must be answered in the negative, at least in regard to normal and not very intense narcosis, and this can be demonstrated experimentally for the osmotic interchange of the cells with the surrounding medium.* Moreover, with a normal, non-damaging narcosis, no significant disturbance could have occurred with respect to the internal osmotic cell exchange, i.e. the osmotic exchange between the various cell organelles, since the consequences of such a disturbance would not immediately disappear once the narcotic is removed from the cell. For very intense narcoses producing cell disturbances that remain for some time following removal of the narcotizing compound, the possibility of a slight change in the osmotic properties of cells during narcosis is not excluded, and may even be likely. However, such changes are less likely to take place with excessively strong brain narcoses than with general cell narcosis, i.e. in the case of excessively strong narcosis of

* This author considers as osmotic processes only those phenomena related to the absorption and loss of substances which are controlled by diffusion processes and by the distribution of compounds absorbed to specific solvents within the cells, and not the exchange of substances between cells and their medium that is dependent upon a particular activity of the protoplasm.

plant cells, muscle fibres, etc. In most cases, it is difficult to determine whether a disturbance in osmotic properties of relevant cells represents the primary effect of fatal narcosis of plant cells, blood corpuscles, etc., and the death of the cells a secondary effect, or *vice versa*, since both phenomena normally follow one another immediately.

It is highly probable that the functions of brain or plasma lipoids depend not upon their importance to the osmotic properties of the cell, but to other tasks they perform related to cell function. However, until the exact relationships have been discovered between brain lipoids and cell proteins, which will no doubt at the same time reveal the fine structure of the protoplasm, it is impossible to make more than vague conjectures on the further functions of brain lipoids.

We already know that it is possible to interrupt the conduction of stimuli through the axial cylinders by means of local effects of narcotics on the nerve fibres, without permanently destroying their ability to conduct stimuli. For this, of course, much higher concentrations of the narcotics are needed than for narcosis of the ganglia cells. I am very inclined to believe that narcosis of the ganglia cells also depends upon a quite similar increase in resistance that they produce within the path of stimulus conduction. Already, under normal circumstances, the resistance encountered in the conduction of a stimulus appears to be much greater within the ganglia cells, or rather between the ganglia cells, than the conduction resistance in the centripetal and centrifugal nerves. The relatively long reflex time compared with the conduction velocity in the nerve stem points to this, at least, as well as the fact that it requires a much stronger action to stimulate a reflex movement than by direct stimulation of the motor nerves. It does not seem at all remarkable to me that the action of narcotics should be restricted at first primarily to an increase in this intra- and intercellular resistance, and that only in higher concentrations do narcotics exert any perceptible effect on stimulus conduction in the nerve fibres. Furthermore, the transmission of the stimulus from the sensory to motor pathways in the central part of the reflex arc is certainly not merely stimulus conduction. At any rate, processes that take place in the intermediary ganglia cells serve as new sources of stimulation. It is likely that it is the presence of narcotics in the ganglia cells and their branches, the neurons and dendrites, that provides greater resistance to the performance of these processes. It is not of value at this time to pursue how the state of brain lipoids that is altered by the absorption of non-specific narcotics evokes this greater resistance.

The basic narcotics also act, presumedly, by increasing the resistance to stimulus transmission, with the difference that this resistance is produced or increased in a different fashion.

A detailed study of narcosis of the muscle and nerve fibres would certainly be the most appropriate way to gain a more profound insight into the mechanism of narcosis, since changes in the activity of these fibres, i.e. the muscle fibres, are the easiest to measure.

It is of enormous interest to me that ethyl ether and chloroform produce complete narcosis in amphibia and mammals, including humans, at the same concentration in the blood plasma. This will probably also apply more or less to most of the other non-specific narcotics whose effect is purely narcotic. In general, however, the concentrations of narcotics in the blood plasma required for narcosis cannot be expected to be the same in cold- and warm-blooded animals, since the partition coefficients of the narcotics between water (the blood plasma) and the brain lipoids will often change, depending upon the temperature. For chloral hydrate, it can easily be calculated that the concentration required for complete narcosis in mammals cannot differ much from that required for amphibia. About 0.75 grams per kilogram of animal is required for complete narcosis in mammals. Since the partition coefficient of chloral hydrate between water and oil favours water rather strongly, only a small amount of the chloral hydrate passes into the fats of the animal, and we can therefore calculate its concentration by dividing the 0.75 g by the numerical value of the water content of 1 kg of animal. If we assume the water content to be 650 g, then we find that the concentration of chloral hydrate in the fluids of the narcotized mammal is about 1 : 850, whereas in tadpoles, we found 1 : 800–1 : 1200. If there was a need to determine the customary doses of normal hypnotics for humans, one would, of course, obtain much lower values for the concentrations of the narcotics in the fluids than we found were required for complete narcosis in tadpoles, even if the removal of the narcotics from the fluids by the fats is disregarded. Attention must be paid here to two things. First of all, under normal circumstances, the effects of the hypnotics act in concert with those natural factors producing sleep, and secondarily, sleep represents to a certain extent, only a preliminary stage of complete narcosis, since most of the reflexes are retained during sleep.

The simplest way to decide if a narcotic produces complete narcosis at the same concentration in mammals, tadpoles, crustacea, and other organisms would naturally be to introduce, first of all directly into the blood of a medium-sized or fairly large mammal, a sufficient quantity of the narcotic that would just produce narcosis in the animal. One could then draw a reasonably large sample of blood (10–50 ml) and perform with the aid of a centrifuge a separation of blood corpuscles and serum at body temperature. Once this separation had been completed, one could then place tadpoles, crustacea, and other organisms in the serum and observe whether or not they also become completely narcotized in it. I am hoping to conduct a series of such experiments soon, possibly after decreasing the alkalinity of the serum, since tadpoles are very sensitive to alkaline solutions. The dehydrating action of mammalian blood, whose osmotic pressure is much higher than that of amphibian blood, occurs too slowly in most cases to have a disturbing effect.

A knowledge of any chemical change that the narcotics undergo within the organism is essential to a more exact understanding of the course of narcosis induced in mammals by non-specific narcotics that are essentially non-volatile at body temperature. We are already aware of these chemical transformations for many narcotics. Thus, chloral hydrate and the tertiary alcohols are gradually conjugated with glucuronic acid, and the salts of these conjugated acids can be actively excreted through the kidneys. This does not apply to the non-specific narcotics themselves, since they pass into the urine in small amounts as a result of a simple diffusion, but are not controlled by the activity of specialized cells. Pyridine is gradually converted to an ammonium compound. This derivative, or its salts, are also actively excreted by the epithelia of the kidneys. Unfortunately, at the moment, our knowledge of these processes is only of a mostly qualitative nature, while we know very little of the quantitative and time-dependent relationships involved in these conversions.

We know that two non-specific narcotics hardly ever have exactly the same effect, since different nerve constituents undergo narcosis in a somewhat varying order. In many cases, the differences in the effects of different narcotics can no doubt be attributed to side effects, i.e. effects that extend to other components besides the brain lipoids. *A priori*, however, it seems not unlikely that the percentage of individual brain lipoids is not the same in different areas of the central nervous system, which must have caused the various non-specific narcotics to behave differently. One very remarkable fact is that, given two compounds which are chemically closely related, one may act predominately as a narcotic and the other may primarily produce spasms, a phenomenon to which, for example, our attention was directed by Richet, and for which this study provides some examples. Are the brain lipoids of certain ganglia cells also involved in some way in these effects?

Especially interesting is the fact first discovered by Claude Bernard that the decomposition of carbon dioxide in the green part of plants in light can be arrested or at least temporarily reduced by ether or chloroform without killing the plant parts involved, and that also development and growth can be stopped by these compounds. The question whether or not carbon dioxide decomposition can be completely stopped or only greatly reduced without causing death, seems to me to be of secondary importance. In the course of my investigations of osmosis, I have been able to demonstrate that for more than 50 non-specific narcotics and antipyretics, about the same concentration inhibiting the flow of protoplasm in algae also stops their decomposition of carbon dioxide in light, or at least reduces it so much, that it no longer exceeds the amounts of carbon dioxide produced by respiration. Whether or not the respiration quotient in plants is altered during narcosis is a question that cannot be answered with certainty at the moment, since in the experiments conducted on this subject, insufficient attention was paid to the relative concentrations of the narcotics. It is known that in animals the amount of carbon dioxide produced during narcosis is very much reduced. However, in this case, it is not a question of a direct effect of the narcotic (which cannot make itself felt until much higher concentrations are reached) on the elements that are first involved in carbon dioxide formation (muscle fibres and gland cells), but of the loss of the tonic effect of the motor nerves, as is caused by the action of curare. Only in the case of carbon dioxide narcosis would it be possible that the carbon dioxide has a direct effect as a result of the lowering of the alkalinity of the cell tissues and plays a larger role.

Although numerous non-specific narcotics have a fatal effect in sufficient concentrations on many if not most micro-organisms in a state of growth, they are generally not able to permanently destroy the development ability of the spores of these micro-organisms within a reasonable period of time, and therefore, cannot be considered true antiseptics.

The better organic antiseptics are similar to the non-specific narcotics in that, like them, they penetrate extremely quickly into all living plant and animal cells and can pass out of them again. They also pass in some quantity into the brain lipoids, but still possess the characteristic of acting on cell proteins, even at low concentrations. As with narcosis, their influence on the cell proteins involves reversible processes that are quickly counteracted when the concentration of the antiseptic in the blood of the test animal is decreased. If there has been a stronger effect, however, the life of the protoplasts is destroyed. Some good antiseptics that are difficult to dissolve in water and that have a partition coefficient between water and oil that favours the oil, such as, for example, thymol, act on tadpoles and other organisms in fairly dilute solutions exactly like the non-specific narcotics; and in the phenol series, one finds overlapping between antiseptics and non-specific narcotics. It has already been emphasized that there is no sharp dichotomy between antipyretics and the non-specific narcotics, although many antipyretics, such as resorcinol or antipyrine, are not able to produce complete brain narcosis without fatality, since they paralyse the heart. All non-specific narcotics induce a narcosis-like state in plant cells, ciliary cells, etc., but in some cases, such as with resorcinol, this is obviously different from narcosis induced by true non-specific narcotics.

9.1 SUMMARY OF SOME OF THE FINDINGS

1. Narcotics can be subdivided into two main groups, non-specific narcotics and basic narcotics. However, these two groups overlap one another.
2. All narcotics, whether non-specific or basic, penetrate rather readily into undamaged plant and animal cells; the large majority of them do this extraordinarily rapidly. They can be transported out of the living cells just as readily once the concentration of the narcotic in the surrounding medium is reduced.
3. Non-specific narcotics act primarily as follows: they can be transported into the lecithin and cholesterol-related components of cells and in this fashion alter the physical state of these 'brain lipoids' ('plasma lipoids') in such a way that they themselves can no longer fulfil their normal functions within the cell, or they have a disturbing effect upon the functions of other cell components.
4. The narcotic potency of a non-specific narcotic is primarily governed by the value of its partition coefficient between water (or the aqueous fluids of the organism) and the brain lipoids (plasma lipoids) as solvents.
5. Following from proposition 4, the narcotic strength of a compound increases rapidly in the various homologous series with the length of its carbon chain. If, however, as with the highest members of the series, its absolute solubility in the brain lipoids falls below a certain minimum, the compound can no longer act as a narcotic, despite the very high value of its (brain lipoids/water) partition coefficient.
6. Among the isomeric alcohols, esters, etc., the compound with the least branched chain is the most potent narcotic, and the compound with the most branched carbon chain is the weakest narcotic, which also follows from proposition 4.
7. The introduction of one hydroxyl group into a molecule in place of a hydrogen or halogen atom considerably lowers the narcotic strength of a compound. This applies even more when two or more hydroxyl groups are introduced into the molecule. However, if the hydrogen of the hydroxyl groups is replaced by an alkyl group, the resulting compounds function again as potent narcotics. These changes in the narcotic potency of a compound after various substitutions also follow from proposition 4.
8. The most potent narcotics are those compounds which have both a

very low water solubility and a very high solubility in ether, olive oil or the 'brain lipoids'. For example, phenanthrene, which is soluble only in about 200 000 parts of water, but is very soluble in olive oil, produces narcosis in tadpoles at a concentration of 1 : 1 500 000 (chloroform does not narcotize until it reaches a concentration of 1 : 6000).

9. Ether and chloroform produce narcosis in humans, mammals, tadpoles, and entomostraca at about the same concentration in the blood plasma (ether 1 : 400, chloroform 1 : 4500–1 : 6000). The same is probably more or less correct for other pure-acting (i.e. free of side effects) non-specific narcotics. On the other hand, the various groups of worms, for the most part, are not narcotized until double and triple the concentration of non-specific narcotics is reached with respect to that needed for narcosis in tadpoles. Usually six to ten times higher concentrations of non-specific narcotics are needed to narcotize plant cells, protozoa, ciliary cells, etc. than are needed for narcosis in tadpoles. These differences in concentrations are much lower only in the case of the non-specific narcotics that are more soluble in water than in olive oil, e.g. methyl and ethyl alcohol or acetone.
10. Under normal conditions, amphibia, insects, etc. become completely narcotized at a much lower partial pressure of the volatile non-specific narcotics than in the air inhaled by mammals or birds. This phenomenon results from the fact that at a given partial pressure of the narcotic, the blood absorbs much greater quantities of the narcotic at a lower temperature than at a higher temperature.
11. The organic antiseptics (carbolic acid, the cresols, thymol, etc.) are like the non-specific narcotics in that they pass extremely readily in and out of living cells. They also pass to some extent into the brain lipoids, but they also have the capacity to form complexes with cell proteins. Every conceivable connecting link exists between non-specific narcotics and antiseptics. In many cases, it is simply a matter of concentration that determines whether a compound should be designated a narcotic or an antiseptic. The antipyretics and non-specific narcotics also overlap one another.
12. Typical basic narcotics (and alkaloids in general) appear to form salt-related complexes with cell proteins, which are in a state of dissociation and become further dissociated as a result of decreasing the concentration of the corresponding free base in the blood plasma. In general, however, when detoxification occurs, a state of equilibrium is reached much more slowly than with the non-specific narcotics. Most basic compounds which are only slightly alkaline act (at least on tadpoles) primarily in the same way as non-specific narcotics.

Appendix A
Detoxification by means of dialysis

Given the fact that almost all non-specific narcotics (chloralose is an exception), antipyretics and antiseptics permeate living cells extraordinarily quickly in both directions, in accordance with the laws of diffusion, without any particular absorbing or secreting activity of the protoplasm being involved, the question arose in my mind some time ago whether it might not be possible to develop a general method of artificially detoxifying an organism that has been poisoned by these compounds.

It is appropriate to begin our discussion of this question with our experiences based upon gill-breathing animals, i.e. tadpoles. It has been demonstrated in numerous experiments that the symptoms of intoxication that tadpoles exhibit in solutions of these non-specific narcotics, once the concentration exceeds a certain level that is characteristic for each individual compound, usually disappear after only a few minutes when the tadpoles are transferred to fresh water. This reversal and disappearance of intoxication symptoms can take longer only in those cases in which the circulation had already become very weak before transfer to toxicant-free water or had almost ceased, or where the partition coefficient of the corresponding compound is extraordinarily high, as in the case of phenanthrene.

The intoxication and detoxification of tadpoles can be repeated with most of these compounds 20, 30 or 50 times in one day without causing the death of the tadpoles, by alternately placing the tadpoles in solutions of the corresponding compounds and in pure water.

It immediately becomes obvious that it must be quite immaterial whether the toxic compound entered the tadpoles through the gills or whether it is introduced into the tadpole's body through the stomach or by peritoneal injection or subcutaneous injection, so long as the concentration of the toxicant reaches the same level in the blood, and we disregard the local effects of the compound that do not concern us here. Therefore, after introduction of the toxic substances into the tadpoles by one of the above mentioned methods, these compounds will pass through the epithelia of the gills and skin into the surrounding medium, so long as this medium is free of the compounds, or contains them in lower concentration than the blood plasma. In the first case, the removal of the foreign compounds from the bodies of the tadpoles will be essentially complete in a short time if either the volume of the medium surrounding the tadpoles is sufficiently large or if, with a smaller volume, the medium is constantly renewed.

If, therefore, gill-breathing test animals are placed in a large container of water or in a small container in which the water is constantly changed, it is possible, without causing the death of the organisms, to inject into the water within one day, in interrupted doses, 12, 20 or 50 times the amount of a non-specific narcotic that would be sufficient to destroy life if the compound did not pass through the gill epithelia, etc. into the surrounding water. Is it possible to artificially accomplish a similar removal of these compounds from the tissue of animals that breathe through their lungs?

This question can at least partially be answered in the affirmative. In naked amphibia, the epidermis of washed animals, although it consists of about seven cell layers (frog) of which the outer layer is horny, is fairly permeable in both directions to almost all compounds (not in both directions for water) which are permitted to pass through all other living plant and animal cells. Accordingly, a frog or triton that has been narcotized by a non-specific narcotic that is not appreciably volatile at room temperature recovers much more quickly if it is placed in a fairly large flat vessel containing water at a level of 2–3 cm than if it is merely held in moist air. However, since the skin of adult amphibia (as a result of the partial cornification of the epidermis) is still for the most part relatively impermeable to non-specific narcotics, detoxification by this method occurs very slowly.

Detoxification can be accomplished much more rapidly if the abdomen of the frog or the triton is opened and the intestines and colon flushed with a constantly renewed solution of 6 parts per thousand sodium chloride. It is also expedient to add a trace of potassium chloride and secondary sodium carbonate (1 : 5000) to the salt solution. The more directly this solution comes into contact with the largest possible surface area of the intestines, peritoneum and the mesenteron, allowing diffusion exchange to take place, and the fewer gaps there are in renewing the solution, the faster the detoxification will take place. For some years already I have successfully been conducting these types of detoxification experiments with frogs and tritons.

In order to be quite clear about this kind of procedure from a theoretical point of view, it is necessary to keep two things in mind. First, it should be remembered that it is almost exclusively the concentration of a narcotic in the blood plasma that determines the onset of brain narcosis, and that the length of time that it acts has almost no influence. Thus, for example, concentration c of a narcotic in the blood plasma can produce complete brain narcosis within 2–3 minutes or even in fractions of a minute, while a concentration of $\frac{2}{3}c$ of the same narcotic does not produce complete narcosis even after several days. Nevertheless, the duration of the toxicant's action appears to be a determining factor in the intensity of some other functional disorders. In the discussion of the experiments with ethyl ether, it was particularly emphasized that circulation can continue in tadpoles for a considerable length of time even if they are kept in a solution that has an ether concentration that is more than double the concentration required for complete narcosis. In the same place, it was reported that similar relationships occur in the action of many other non-specific narcotics. The longer the experiment is continued, however, the weaker becomes the activity of the heart, although the concentrations of ether in the blood and in the tissues have long before been equalized. Exactly the same thing applies to the effects of ether and many other narcotics on the respiratory centre, which usually tends to become paralysed by narcotics sooner than the heart.

Therefore, if an amphibian that has received an otherwise fatal dose of a non-specific narcotic is partially rid of this compound soon after it is administered by means of the method described above, it can frequently survive (the same applies to many other compounds that readily penetrate living cells).

Secondly, it must be kept in mind that, if a toxicant, e.g., chloral hydrate, gets into the organism through the intestinal tract, as is usually the case, the toxicant passes into the blood very gradually, even if the living cells are very permeable to the compound. The larger the animal, the slower, in general, will be the absorption of the compound into the blood, since the ratio of the surface area of the stomach to its volume will be reduced and, perhaps even more significantly, the distance from the axial parts of the stomach from its walls is increased. This last factor naturally produces a reduced steepness in the concentration gradient of the toxic compound that results from its absorption across the mucous membrane. After the toxicant has passed out of the stomach into the intestinal canal, these same relationships still play a certain role, that is if the chyme is not too watery. Washing out the peritoneal cavity with the salt solution under these conditions would therefore frequently permit the toxic concentration in the blood to be kept below the fatal level. The threatened increase in toxic concentration in the blood resulting from the continued absorption of the toxicant from the intestinal tract is abolished by the simultaneous removal of the toxicant from the circulation as a result of its transfer into the salt solution.

The further question now arises naturally whether mammals and humans can also be detoxified in this same way, as with amphibians. As far as the possibility of detoxification by means of deep baths is first of all concerned, it can be predicted with some certainty that this would achieve very little, since although the epidermis of living mammals is, in fact, fairly permeable to quite a number of compounds (phenol, thymol, alcohol, etc.), for the most part the toxicants would pass much too slowly into the water bath to be of any practical value. Nonetheless, it is conceivable that in a few cases, a very warm bath, which, by the way, can be useful for other reasons in certain types of intoxication, might cause a fairly substantial amount of the toxicant to pass into the bath water. This is because, under these conditions, the blood flow to the skin is very much increased and also the activity of the sweat glands is greatly increased.

For mammals and humans, passing a warm salt solution through the peritoneal cavity is more likely to be of value. Experiments in this direction, conducted in collaboration with Professor M. von Frey during the Easter vacation in 1899, using rabbits that had been intoxicated with chloral hydrate, were at least successful enough to warrant additional experiments. Chloral hydrate was chosen solely for its practical interest. Otherwise, this particular compound is not very suitable for experiments of this type. When doses are administered that would be great enough under normal conditions certainly to cause death, there is a tendency for death to occur so quickly that there is only time for a small amount of the salt solution to be passed through the peritoneal cavity. At least this is true if the chloral hydrate is injected into the rectum, from which it is absorbed completely much more rapidly and therefore has a much more intensive effect than when injected subcutaneously. Rabbits can sometimes survive doses of 2 grams of chloral hydrate given by subcutaneous injection, without any active intervention. If the same amount of compound is injected into the rectum, death frequently occurs within 8 minutes. In the case of chloral hydrate, a self-intoxication occurs as a result of the gradual conversion of the already absorbed chloral hydrate into urochloral acid, if the chloral hydrate is not absorbed too quickly into the blood.

It is certain that if a salt solution is passed through the abdominal cavity, the symptoms of intoxication certainly disappear more quickly than they would have without this intervention. In one experiment, $1\frac{3}{4}$ grams of chloral hydrate was administered rectally in a single dose to a rabbit weighing $2\frac{1}{8}$ kilograms, and then about $4\frac{1}{2}$ litres of 0.85% sodium chloride solution that had been heated to 38°C and, of course, sterilized first, was passed through the peritoneal cavity. In this process, the abdominal cavity was alternately filled up more with the salt solution and was then partially emptied again by gently pressing the stomach walls, so that the newly entering (still free of chloral hydrate) salt solution came into contact with a larger surface area of the intestines, mesenteron, etc. The reflexes returned after only 45 minutes, and the rabbit subsequently recovered completely. It might be recommended, by the way, to use in place of a simple 0.85% salt solution, a solution containing approximately 0.7% NaCl, and also about 0.07% each of Na_2CO_3 and $NaHCO_3$ and a trace of KCl, so that the natural proportions of the salt content of the body fluids are more closely approximated.

For all such toxicants that pass rapidly by purely diffusion out of the blood into all the different cell tissues and from the cell tissues back into the blood, according to the direction of the drop in concentration at that moment (or the decrease in pressure of the relevant compound) it seems much more sensible, at least from the physiological standpoint, experimentally to remove the toxicant from the organism by the above described method than to do it by removing a certain quantity of blood from the intoxicated organism and replacing it with a physiological salt solution.* Only in those cases in which the toxicant exerts primarily a damaging effect on the blood itself, possibly by decomposing the blood corpuscles, does the latter method seem to me to be justified. The weakening of the organism by the removal of the organic constituents of the blood would, in most cases, cause more harm than the only slight decrease in toxic concentration achieved by this method would benefit the organism. In the case of the toxicants being studied, at any given moment a far larger proportion of the toxicant is present in the tissues than the blood. The method devised by Sahli [1], whereby a salt solution is introduced into the subcutaneous tissue, thereby stimulating kidney action, can also cause only relatively small amounts of the toxicant to be removed from the organism. These toxicants pass only by diffusion (and

* It was shown in the discussion of ether narcosis (pp. 101–4) that a dissolved compound can move from a point of higher concentration to one of lower concentration if the medium in which it is contained has an equal dissolving capacity for this compound (e.g. a concentrated and a more dilute salt solution). Therefore, it seems appropriate to introduce, for all compounds in solution, the concept of its pressure in a corresponding sense as the latter has been used for some time in physiology for solutions of gases. The direction in which these pressures decrease would then define the direction of the diffusion flow of the compound. (Editor's note: this concept presages the application of chemical potential to toxicity much later by Ferguson, J. (1939) *Proc. R. Soc.* (London), Ser. B, **127**, 387.)

not by an active secretion of the epithelia of the kidneys) into the urine, as is in fact the case with non-specific narcotics so long as they are not chemically changed by the organism (e.g. conjugated with glucuronic acid).

Of course it is possible with all such compounds, that merely obey the laws of diffusion in their distribution throughout an animal's body, to remove a certain amount from the organism by passing a constant flow of liquid through the stomach or alternately filling and emptying the stomach with water or a salt solution. So long as the blood plasma toxicant concentration exceeds that of the stomach, the diffusion flow will be directed against the lumen of the stomach. By using a hot (38–40°C) solution for this flushing, and possibly adding certain drugs to the solution, a more vigorous circulation can be produced in the stomach wall, which would of course be advantageous for this type of detoxification.

In the specific case of intoxication with alkaloids, it would seem very sensible to pass a solution of Glauber's salt, which has been slightly acidified with sulphuric acid, through the stomach for the following reasons.

It has been known for some time that, if a solution of a morphine salt is administered subcutaneously to an organism, a considerable amount of the morphine is passed into the stomach. Exactly the same thing happens, however, in the case of many other alkaloids, probably in every case. This transfer of morphine into the stomach is usually explained as an active secretion of the morphine by the mucous membrane of the stomach, but this assumption is hardly necessary, for even in the absence of such an active secretion, the alkaloids would very clearly have to pass out of the blood into the stomach so long as the stomach contents produce an acid reaction.

For the most part, alkaloid salts are decomposed by the alkali of the blood, and the free alkaloids, which, in contrast to the alkaloid salts, are only slightly ionized, readily penetrate all living protoplasts in both directions. Therefore, the free alkaloids, following the laws of diffusion, must also partly penetrate through the epithelia of the stomach. If, however, a free acid is present in the stomach, the alkaloids become bound to this acid, i.e. ionized to a large extent, and the conditions are therefore present for a further transfer of the free alkaloid from the blood into the stomach. In exactly the same fashion, alkaloids formed within the alkaline-reacting protoplasm of plant cells accumulate in the acid-reacting cell fluid. In both cases, the possibility of the newly formed alkaloid salts returning from the stomach into the blood, or from the acid cell fluid into the protoplasm of the plant cells, is excluded in the absence of an active protoplasmic action, since living cells are completely, or almost entirely impermeable to alkaloid salts as such. A diffusion of alkaloid salts through the putty-like substance of the stomach epithelia would also perhaps be possible, but by no means all putty-like substances are permeable to salts.

Under normal circumstances, the alkaloid salts formed in the stomach are again dissociated by the alkali in the intestinal juice after they pass into the intestinal tract, and they return to the blood. Perhaps the non-dissociated salts are also partly absorbed as a result of an active absorption on the part of the intestinal epithelia, or they diffuse through the putty-like substance of the epithelia. Thus, in this case, the alkaloids are maintained in a continuous cycle that goes from the intestinal tract through the blood to the stomach and from the stomach back to the intestine, until they are finally destroyed, chemically converted, or removed from the organism through the kidneys. If however, a flow of acidifed water or an acidified salt solution is passed through the stomach, or if the stomach is alternately filled with an acidified solution and then emptied, a significant portion of the alkaloid can be removed from the organism. The use of a solution of Glauber's salt may more or less prevent an absorption of the portion of the stomach contents that pass into the intestinal tract during this procedure. An abundant excretion of very acid urine would also certainly promote the removal of an alkaloid from the organism.

REFERENCE

1. Sahli (1890) Ueber Auswaschung der menschlichen Organismus, etc. *Correspondenz-Blatt für Schweiz. Aerzte,* 20th annual edn.

Appendix B
List of publications of Charles Ernest Overton

(1888) Ueber den Conjugationsvorgang bei *Spirogyra*. *Ber. Dtsch. Bot. Ges.*, **6**, 68–72.

(1889) Beiträge zur Kenntnis der Gattung *Volvox*. (Inaugural Dissertation). *Bot. Centralbl.*, **39**, 65–72, 113–8, 145–50, 177–82, 209–14, 241–6, 273–7.

(1891) Beiträge zur Histologie und Physiologie der Characeen (Habilit–Schr.). *Bot. Centralbl.*, **44**, 1–10, 33–8. Editor's note: these two articles have the same title and appear in **44** (1) and (2).

(1890) Mikrotechnische Mitteilungen aus dem botanischen Laboratorium der Universität Zürich. *Zeitschr. wiss. Mikrosk.*, **7**, 9–16.

(1891) Beitrag zur Kenntnis der Entwicklung und Vereinigung der Geschlechtsprodukte bei *Lilium Martagon*. In: *Festschr. zur Feier des 50 jährigen Doctorjubiläums der Herren Prof. v. Nägeli und Prof. v. Kölliker*, Zürich, 11 pp.

(1893) On the reduction of chromosomes in the nuclei of plants. *Annals Bot.*, **7**, 139–43.

(1893) Ueber den gegenwärtigen Stand der Befruchtungslehre bei den Pflanzen, *Ber. Schweiz. Bot. Ges.*, **3**, 11–12.

(1893) Ueber die Reduction der Chromosomen in den Kernen der Pflanzen. *Vierteljahrsschr. Naturforsch. Ges. Zürich*, **38**, 169–86.

(1894) *Ber. Schweiz. Bot. Ges.*, **4**, 18.

(1895) Über die osmotischen Eigenschaften der lebenden Pflanzen- und Tierzelle, *Vierteljahrsschr. Naturforsch. Ges. Zürich*, **40**, 159–201.

(1896) Über die osmotischen Eigenschaften der Zelle in ihrer Bedeutung für die Toxikologie und Pharmakologie. *Vierteljahrsschr. Naturforsch. Ges. Zürich*, **41**, 383–406; also reprinted in (1897) *Zeitschr. physik. Chem.*, **22**, 189–209.

(1897) Notizen über die Grünalgen des Ober-Engadins. *Ber. Schweiz. Bot. Ges*, **7**, 49–68.

(1897) Ueber zwei für die Schweiz neue Algenarten. *Ber. Schweiz. Bot. Ges.*, **7**, 6–7.

(1899) Über die allgemeinen osmotischen Eigenschaften der Zelle, ihre vermutlichen Ursachen und ihre Bedeutung für die Physiologie. *Viertel-jahrsschr. Naturforsch. Ges. Zürich,* **44**, 87–136. An English translation of a large excerpt from this article may be found in: On the general osmotic properties of the cell, their probable origin and their significance for physiology. *Cell Membrane Permeability and Transport,* (ed. G.R. Kepner), Dowden, Hutchinson and Ross Inc., Stroudsburg, PA, pp. 29–56.

(1899) Notizen über die Wassergewächse des Oberengadins. *Vierteljahrsschr. Naturforsch. Ges. Zürich,* **44**, 211–28.

(1899) Experiments on the autumn colouring of plants, *Nature,* **59**, 296.

(1899) Beobachtungen und Versuche über das Auftreten von rotem Zellsaft bei Pflanzen. *Jahrb. wiss. Bot.,* **33**, 171–231.

(1900) On the osmotic properties and their causes in the living plant and animal cell. *Brit. Ass. Rep.,* 940–1.

(1900) Studien über die Aufnahme der Anilinfarben durch die lebende Zelle. *Jahr. wiss. Bot.,* **34**, 669–701.

(1901) *Studien über die Narkose, zugleich ein Beitrag zur allgemeinen Pharmakologie,* G. Fischer, Jena, 195pp.

(1902) Beiträge zur allgemeinen Muskel- und Nervenphysiologie. I. Mitt. Über die osmotischen Eigenschaften der Muskeln, *Pflügers Archiv ges. Physiol.,* **92**, 115–280.

(1902) Beiträge zur allgemeinen Muskel- und Nervenphysiologie. II. Mitt. Über die Unentbehrlichkeit von Natrium (oder Lithium) Ionen für den Kontraktionsakt des Muskels, *Pflügers Archiv ges. Physiol.,* **92**, 345–86.

(1903) Über die Unentbehrlichkeit der Natrium-ionen für die Tätigkeit des zentralen und peripheren Nervensystems. *Verh. Ges. Dtsch. Natursforsch. u. Ärzte* **75**, (2) 416–9.

(1904) Beiträge zur allgemeinen Muskel - und Nervenphysiologie. III. Mitt. Studien über die Wirkung der Alkali- und Erdalkalisalze auf Skelettmuskeln und Nerven, *Pflügers Archiv ges. Physiol.,* **95**, 175–290.

(1904) Neununddreissig Thesen über die Wasserökonomie der Amphibien und die osmotischen Eigenschaften der Amphibienhaut. *Vehr. physik.-med. Ges. Würzburg, N.F.,* **36**, 277–95.

(1905) Über reversible Änderungen in der Spannung und Richtung des Demarkationsstromes nach Ersatz der Gewebelymphe der Muskeln durch andere Lösungen. *Sitz. Ber. Physik.-med. Ges. Würzburg,* 2–7.

(1907) Über den Mechanismus der Resorption und Sekretion In: *Handbuch der Physiologie des Menschen,* W. Nagel, ed., **2**, 744–898.

(1907) Ueber die Abhängigkeit der Spannung und Richtung des Demarkations-stromes von der Beschaffenheit der die lebenden Muskelfasern umspülenden Losüngen. *7 Internat. Physiol.-Kongress,* Heidelberg.

(1907) Demonstration von Apparaten zu quantitativen Versuchen über die Quellung, *7 Internat. Physiol.-Kongress,* Heidelberg.

(With I. Bang) (1911) Studien über die Wirkungen des Kobragiftes. *Biochem. Zeitschr.,* **31**, 243–93.

(With I. Bang) (1911) Studien über die Wirkungen des Crotalusgiftes. *Biochem. Zeitschr.,* **34**, 428–61.

(1913) Studien über einige Wirkungen der Saponine. Lunds Universitets Årsskrift, N.F. Avd. 2, 9, 7, 27 pp.
(1918) Några ord om våra träds rikliga blomning innervarande år samt om bokollonens betydelse i näringsfysiologiskt hänseende, *Hyg. Revy.*, **7**, 77–9.
(1918) Untersungungen über the Resportion und die relative Stärke einiger Herzgifte, nebst einleitenden Versuchen mit Salzen der Alkalien und Erdalkalien. Lunds Universitets Årsskrift, N.F. Avd. 2, **14**, 7, 48pp.
(1925) Über den Mechanismus der Aufnahme und über das Verhalten der Ester im Organismus. *Skand. Arch. Physiol.*, **46**, 333–4.
(1925) Eine reversible, resorptive Lähmung der motorischen Nervenenden beim Frosche durch gewisse indifferente Narcotica. *Skand. Arch. Physiol.*, **46**, 335.
(1926) On a method of rendering the skin of the living frog permeable to salts, carbohydrates and other crystalloids. *Skand. Arch. Physiol.*, **49**, 196.

Appendix C
Publications about Overton and analyses of his data

Collander, R. (1959) in *Plant Physiology: A Treatise,* (ed. F.C. Steward) Vol. 2, Academic Press, New York, pp. 7–9.

Collander, P.R. (1962–3) Ernest Overton (1865–1933): a Pioneer to Remember. *Leopoldina,* **8–9**, 242–54.

Die Universität Zürich 1833–1933 und Ihre Vorläufer. Festschrift zur Jahrhundertfeier. (eds E. Gagliardi, H. Nabholz, and J. Strohl) (1938) Verlag der Erziehungsdirection, Zürich, pp. 871, 872, 948, 998.

Gillespie, C.C. (1974) Charles Ernest Overton. *Dictionary of Scientific Biography,* Vol. X, Charles Scribners, New York, pp. 256–7.

Hansch, C. and Dunn, W.J. (1972) Linear relationships between lipophilic character and biological activity of drugs. *J. Pharm. Sci.,* **61**, 1–19.

Holmstedt, B. and Liljestrand, G., (eds) (1981) Charles Ernest Overton, in *Readings in Pharmacology,* Raven Press, New York, pp. 150–4.

Jacobaeus, H.C. (1935) Charles Ernest Overton. *Kungl. Svenska Vet.-Akad. Arsbok,* 261–6.

Kniffki, K.-D. and Westphal, W. (1988) *Würzburger Physiologen: Wissenschaftliches Persönliches,* Nachdenkliches, Medikon Verlag, Munich, pp. 37–65.

Leo, A., Hansch, C. and Church, C. (1969) Comparison of parameters currently used in the study of structure–activity relationships. *J. Med. Chem.,* **12**, 766–71.

Lipnick, R.L. (1986) Charles Ernest Overton: narcosis studies and a contribution to general pharmacology, *Trends Pharmacol. Sci.,* **81**, 161–4.

Lipnick, R.L. (1989) A quantitative structure–activity relationship study of Overton's data on the narcosis and toxicity of organic compounds to the tadpole, *Rana temporaria*. In *Aquatic Toxicology and Environmental Fate: Eleventh*

Volume, ASTM STP 1007 (eds G.W. Suter II and M.A. Lewis) American Society for Testing and Materials, Philadelphia, pp. 468–89.

Lipnick, R.L. (1989) A QSAR study of Overton's tadpole data. In *QSAR: Quantitative Structure-Activity Relationships in Drug Design*, (ed. J.L. Fauchère) (1988) Proceedings of the 7th European Symposium on QSAR, Interlaken, Switzerland, September 5–9, Alan R. Liss, New York, pp. 421–4.

Lipnick, R.L. (1990) Selectivity, in Vol. 1 *Comprehensive Medicinal Chemistry*, C. Hansch, Chairman, Pergamon Press, Oxford, pp. 239–47.

Lorenc, V. and Laboulais, J. (1930) *Les Poisons Overtoniens*, Éditions Medicales Norbert Maloine, Paris.

Malm, A. and Wilner, P. (1924–5) Lunds Universitets Martikel, C.W.K. Gleerup, Lund, Sweden, pp. 226–8.

Santesson, C.G. (1933) Charles Ernest Overton. *Hygiea*, **95**, 161–7.

Siwe, A.S.A. (1918) (Sept). Professor E. Overtons Förepäsmägar Farmakologi, Lunds Skrivbykä, Lund (unpublished lecture notes, Pharmacological Institute, University of Lund).

Thunberg, T. (1933) Ernest Overton. *Proc. Roy. Physiogr. Soc. Lund*, **3**, 45–52.

Thesleff, S. (1964) Ernest Overtons forskargärning. *Nordisk Med.*, **71**, 549–53.

Url, W. (1976) Charles Ernest Overton. 75 Jahre Lipoidtheorie. *Vehr. Zool. Bot. Ges. Wien*, **115**, 24–33.

Index

Page numbers in **bold** refer to tables, *italics* refer to figures and in superscript refer to footnotes and Chemical Abstract Registry Numbers in brackets.

MIX
Papier aus verantwortungsvollen Quellen
Paper from responsible sources
FSC® C105338

If you have any concerns about our products,
you can contact us on
ProductSafety@springernature.com

In case Publisher is established outside the EU,
the EU authorized representative is:
Springer Nature Customer Service Center GmbH
Europaplatz 3, 69115 Heidelberg, Germany

Printed by Libri Plureos GmbH
in Hamburg, Germany